UNSCALED

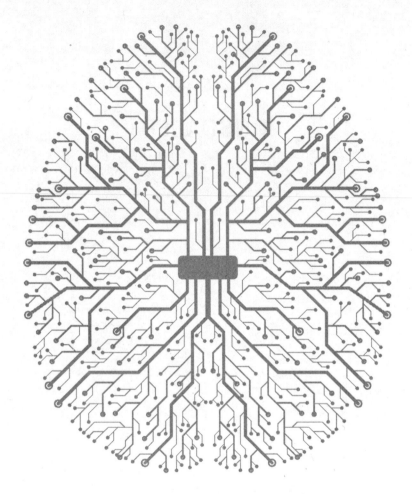

UNSCALED

How AI and a
New Generation of Upstarts
Are Creating the
Economy of the Future

HEMANT TANEJA
WITH KEVIN MANEY

PUBLICAFFAIRS

New York

PublicAffairs
Hachette Book Group
1290 Avenue of the Americas, New York, NY 10104
www.publicaffairsbooks.com
@Public_Affairs

Printed in the United States of America
First Edition: March 2018

Published by PublicAffairs, an imprint of Perseus Books, LLC, a subsidiary of Hachette Book Group, Inc. The PublicAffairs name and logo is a trademark of the Hachette Book Group.

The Hachette Speakers Bureau provides a wide range of authors for speaking events. To find out more, go to www.hachettespeakersbureau.com or call (866) 376-6591.

The publisher is not responsible for websites (or their content) that are not owned by the publisher.

Print book interior design by Cynthia Young at www.sagecraftbookdesign.com

Library of Congress Cataloging-in-Publication Data
Names: Taneja, Hemant, author.
Title: Unscaled : how AI and a new generation of upstarts are creating the economy
    of the future / Hemant Taneja with Kevin Maney.
Description: First Edition. | New York : PublicAffairs, 2018. | Includes bibliographical
    references and index.
Identifiers: LCCN 2017042248 | ISBN 9781610398121 (hardback) |
    ISBN 9781610398138 (ebook)
Subjects: LCSH: Automation—Economic aspects. | Artificial intelligence—Economic
    aspects. | BISAC: BUSINESS & ECONOMICS / Entrepreneurship. | BUSINESS
    and ECONOMICS / Economic Conditions.
Classification: LCC HC79.A9 T36 2018 | DDC 338/.064—dc23
LC record available at https://lccn.loc.gov/2017042248

ISBNs: 978-1-61039-812-1 (hardcover), 978-1-61039-813-8 (ebook)

LSC-H

10 9 8 7 6 5 4 3 2 1

To my partners at General Catalyst

Contents

PART 1

BIRTH OF A NEW ERA, RIGHT NOW

1

The Great Unscaling

Throughout the twentieth century, technology and economics drove a dominant logic: bigger was almost always better. Around the world the goal was to build bigger corporations, bigger hospitals, bigger governments, bigger schools and banks and farms and electric grids and media conglomerates. It was smart to scale up—to take advantage of classic economies of scale.

In the twenty-first century, technology and economics are driving the opposite—an *unscaling* of business and society. This is far more profound than just startups disrupting established firms. The dynamic is in the process of unraveling all the previous century's scale into hyperfocused markets. Artificial intelligence (AI) and a wave of AI-propelled technologies are allowing innovators to effectively compete against economies of scale with what I call the *economies of unscale*. This huge shift is remaking massive, deeply rooted industries such as energy, transportation, and healthcare, opening up fantastic possibilities for entrepreneurs, imaginative companies, and resourceful individuals.

If you feel that work, life, and politics are in disarray, this transformation is why. We are experiencing change unlike any since around 1900, when, as I will detail later, a wave of new technologies, including the car, electricity, and telecommunication, transformed work and life.

Right now we are living through a similar ground-shaking tech wave, as AI, genomics, robotics, and 3D printing charge into our lives. Artificial intelligence is the primary driver, changing almost everything, much like electricity did more than one hundred years ago. We are witnessing the birth of the AI century.

In an economy driven by AI and digital technology, small, focused, and nimble companies can leverage technology platforms to effectively compete against big, mass-market entities. The small can do this because they can *rent* scale that companies used to need to build. The small can rent computing in the cloud, rent access to consumers on social media, rent production from contract manufacturers all over the world—and they can use artificial intelligence to automate many tasks that used to require expensive investments in equipment and people.

Because AI is software that learns, it can learn about individual customers, allowing companies built on rentable tech platforms to easily and profitably make products that address very narrow, passionate markets—even markets of one. The old mass markets are giving way to micromarkets. This is the essence of unscaling: technology is devaluing mass production and mass marketing and empowering customized microproduction and finely targeted marketing.

The old strategy of beating competitors by *owning* scale has in many cases become a liability and burden. Procter & Gamble, with all its magnificent resources, finds itself vulnerable to a newcomer like the Dollar Shave Club, which can rent much of its capabilities, get to market quickly, target a narrow market segment, and change course easily if necessary. General Motors finds itself chasing Tesla. Giant hospital chains don't know how to respond to AI-driven apps that target patients with a specific condition such as diabetes. The economies of unscale are turning into a competitive edge.

In my work investing in startups as a venture capitalist, unscaling has become my central investment philosophy. I fund or help build companies that can take advantage of AI and other compelling new technologies such as robotics and genomics to peel away business and customers from scaled-up incumbents. By adhering to the philosophy of unscale, our firm has invested early in groundbreaking companies such as Snap, Stripe, Airbnb, Warby Parker, and The Honest

Company. Unscaling has also led me to help nonprofit organizations such as Khan Academy and Advanced Energy Economy, which are reimagining the institutions of education and electric utilities, respectively. My activities have given me both a broad and a deep view into unscaling, helping me to see the big picture.

The story of one of my companies, Livongo, provides insight into the dynamics set in motion by AI and unscaling. Livongo ("life on the go") points to the way unscaling can drive *down* the costs of healthcare while increasing effectiveness. The United States spends more on healthcare than any other nation—$3.5 trillion annually, about 18 percent of gross domestic product. Citizens, corporate leaders, and politicians desperately want to get those costs under control but don't want to lose any quality of our healthcare. AI and unscaling can help in part by turning healthcare more toward personalized medicine that can help prevent more people from getting sick in the first place.

In 2014 I helped Glen Tullman get Livongo off the ground, and he's been the driving force leading the company as its CEO. Tullman was born near Chicago and in college studied economics and anthropology, a somewhat unusual background for a CEO who has made his mark in technology. After completing his education Tullman went on to run a couple of software companies and then in 1997 landed the job of turning around a struggling company called Allscripts. Founded in 1982 as Medic Computer Systems, Allscripts had bounced around for more than a decade as a maker of software for medical practices. Tullman and his team refocused Allscripts on software that lets physicians securely write prescriptions electronically. After spending two years improving Allscripts, Tullman took the company public at a $2 billion valuation, remaining as CEO until 2012.

I got to know Tullman when we both independently invested in a medical data analytics company called Humedica, which United-Health Group bought in 2013. After the sale I wanted to continue working with Tullman, so we started hunting for an idea in the healthcare space.

Tullman was particularly interested in diabetes. For starters, it's the fastest-growing disease in the world, and there are over 30 million people with diabetes in the United States alone. We also knew that diabetes is a manageable disease—people who are careful can live pretty normally. Still, for Tullman the disease is personal. "My youngest son was diagnosed with type 1 diabetes when he was eight," Tullman says. "My mom has type 2. I've been surrounded a better part of my life by diabetes, and I was fascinated by how hard we made it for people to stay healthy." People with diabetes typically need to buy expensive test strips, pricking their fingers several times a day and using the strips to analyze their blood sugar. And then it's up to them to act on the reading. The whole procedure is problematic. The strips are expensive, people don't like to poke themselves, and then, if blood sugar spikes or dives, the person can pass out or have a seizure. Longer term, the disease leads to other comorbidities like retinal blindness, kidney disease, and heart disease because people have a difficult time taking care of themselves effectively.

Tullman and I brainstormed ways to fix this, framing it in the following way: *What if we could figure out a way to eliminate the hassle, to have people with diabetes spend less time, not more time, on their disease, to use all the wonderful innovations that we get from Silicon Valley, but do it in a way that the healthcare system could absorb?* We ran across an inventor who had come up with a wireless glucometer—a way to measure blood sugar using a device that could send the results to medical professionals over wireless networks. Tullman acquired that technology, and we launched Livongo in 2014 to build a service for people with diabetes.

Livongo's approach is simple and focused: it sends you a small mobile device that is both a glucose meter and pedometer (so it can track your exercise). It leverages cellular networks to communicate through the cloud back to Livongo software. As a patient tests glucose levels, and the Livongo device sends back data, Livongo's AI-driven system gets to know that patient. If the system starts to see readings that point to a problem, it sends the patient a message to eat something, or to take a walk, or whatever might help. If the system determines there's a serious problem, the patient gets a call from a health professional within a few minutes of checking his or her blood glucose.

As you might imagine, Tullman signed up his son, Sam, for the service, so the senior Tullman has personal experience. Sam, at this writing, is twenty-one and plays football for the University of Pennsylvania. Tullman recalls when he recently met Sam before one of Penn's football games: "When I got there Sam said, 'Hey Dad, I've got something great to tell you.' I assumed it was about football or sports or girls. Instead he told me, 'I had my first Livongo moment.' I said, 'That sounds good. What does that mean?' Then Sam told me, 'It was four a.m. last night, and I woke up. My blood sugar was thirty-seven.' He's six-foot-three and 240 pounds. He knows he can't even stand up with blood sugar at that level. He said, 'I didn't know what to do. My roommate was out. I didn't know whether to call 911. The phone rang, and it was Kelly.' I said, 'Who's Kelly?' 'She works for you,' Sam said. 'Kelly is one of your CDEs [certified diabetes educators]. When we worked through it, Kelly had me crawl over to the refrigerator. If I passed out, she said that she would call 911, but it all worked out great.' Sam then said, 'I realized you weren't in the business for yourself—you were in the business of making sure people don't feel alone anymore.'"

Livongo created a new way for people to manage diabetes—one that would never have come out of the traditional medical field. It doesn't replace the doctor, but it can help people with diabetes manage their lives so they need far less care from doctors or hospitals, which ultimately saves lots of money for individual patients—and in healthcare spending in society overall. But how is this unscaling?

Over the past four or five decades carbohydrate-heavy diets—pushed by mass-market production and mass marketing of cereals and drinks laced with high-fructose corn syrup—created an epidemic of obesity and, ultimately, diabetes. The medical profession lumped most people with diabetes into one of two categories of the disease—type 1 is genetic and type 2 is diet related—and prescribed a standard treatment. It was a classic mass-market medicine approach. So the healthcare industry scaled up to meet demand. It built diabetes centers and more hospitals and ran every patient, assembly-line style, through the same tests the few times a year they'd be able to visit an endocrinologist, whose schedule was packed. Yet for patients, sugar levels in between appointments can change, rising and falling to dangerous levels, and the disease can progress, adding more costs and more visits to bigger hospitals. People suffering from diabetes end up costing the healthcare system $300 billion a year in the United States alone. (It's only going to get worse globally: within a decade China will likely have more people with diabetes than the entire US population.) The scaled approach can't keep up with the growing number of people with the condition, and it fails to give people with diabetes what they really want: a healthy life.

In reality every person who has diabetes suffers from it differently, and the best way to treat it is different for everybody. So Livongo, a startup, was able to quickly build a product and offer it nationwide—and, eventually, worldwide—by leveraging tech platforms such as smartphones and cloud computing. The software and data from patients allows Livongo to offer more personalized care, making patients feel like they are a market of one, not one insignificant person in a mass market—and that makes for happier customers. The technology allows Livongo to nimbly compete against the diabetes-related offerings of giants such as Johnson & Johnson and UnitedHealth Group, winning a fast-growing subset of their customers and serving them at a profit.

Personalized AI-driven care can reduce the amount Americans spend caring for diabetes by as much as $100 billion just by keeping more people with diabetes well more of the time. Unscaled solutions can change the game and reduce healthcare costs by keeping people well. The nation can save money while at the same time making citizens healthier, happier, and more productive.

Livongo is one small example of what's happening in sector after sector all over the world.

* * *

For more than a century size mattered. Economies of scale reigned as a competitive advantage. They worked like this: if a company spent a billion dollars to develop a physical product and build a factory, the amortized cost, at the extreme, would be a billion dollars to make one unit but only one dollar for each unit if the company produced a billion of them. So scale gave a company a cost advantage over competitors. It also brought other advantages, like an ability to negotiate for lower prices from suppliers and the money to blanket mass media with advertising. Once a company built massive scale and accumulated all its advantages, that scale became a huge barrier against competitors. A newcomer would need to build that scale—at great cost—to effectively take on a highly scaled incumbent.

In many ways scale was a net good for society for a long time. Scale was how the world achieved great things like global banking, air travel, widespread healthcare, and the internet. Scaled industries lifted more people out of poverty in the past fifty years than over the previous five hundred years.

The world we're creating now will work differently. Small entrepreneurial companies routinely befuddle corporate giants. Serving niche markets of passionate customers now beats addressing mass markets of merely satisfied customers—because who wouldn't prefer a product or service tailored just for them? We see this in now-familiar instances like when Uber upended long-established taxi companies or Airbnb out-innovated even smart hotel companies such as Marriott. We've known for a while that big companies and entrenched enterprises, which got accustomed to being business superpowers, need to fear two-person garage startups. But now unscaling is becoming systemic, taking apart whole sectors of the economy. The relationship between scale and success is flipping, as I'll describe throughout this book. The winners will be those who exploit the economies of unscale, not the old economies of scale. This is a trend that began playing out around 2007 and will continue for another two decades.

Whether the kind of world that comes out of unscaling will be beneficial for most people depends on the choices we make, starting now. These will be big and difficult choices about the accountability of technology, the role of education, the nature of work, and even the definition of a person. We'll need to make sure the unscaling revolution benefits society broadly, not just the wealthy or the technologically advanced. Those are huge responsibilities.

Although there are serious issues we must focus on, most of the news about unscaling and the technology behind it is positive. We are opening up new ways to solve some of the world's great problems, including climate change and soaring healthcare costs. If we make the right choices, unscaling can reverse many of the ills mass industrialization has brought on, helping to create a future that works better than the past. But we're just starting on this journey. To predict the full ramifications today of AI and unscaling would be like trying to predict the impact of personal computing back in the 1980s, when Microsoft pitched the then-outrageous idea of a computer on every desk and in every home. Yet unscaling is most certainly our future and the outcome of the development of powerful AI. To overlook or deny this would be irresponsible. Better to understand the coming outsized change, guide it, and reap its rewards.

* * *

The emergence of powerful artificial intelligence and the economic force of unscaling can trace their beginnings to 2007, when the Apple iPhone, Facebook, and Amazon Web Services—pioneering mobile, social, and cloud platforms—took wing at roughly the same time. As more of work and life moved online thanks to such platforms, the amount of data exploded. At first the explosion just seemed like more data that could inform business, and we even called it Big Data, as if that's all there was to it. But Big Data turned out to have a higher purpose. It was the key to making AI, which previously had a long and tortured history of disappointment, into a force that will literally change the world. Other new technologies such as virtual reality, robotics, and genomics are also now breaking out, all driven by the power of AI. (Much more on all that in the next chapter.)

These technologies are becoming the foundations of global platforms. The world has been making platforms for generations—the interstate highway system, the internet, as well as mobile phone networks, cloud computing services, and social networks are all platforms. What is so important about platforms is that they do something so you don't have to. A trucking company, for example, doesn't first need to pave a road to transport a load of beer; an app maker doesn't need to build a mobile network or app store to get its software to consumers. The more platforms we build, the less an individual company—or lone entrepreneur—needs to do by itself in order to create, produce, market, and deliver a product.

Now, for much of the twentieth century, even though some plat-forms, like the highway systems, were in place, most companies still had to build a lot of capabilities by themselves. That need gave rise to the vertically integrated corporation. Vertical integration means own-ing much of the "stack" that gets a product from an idea to a customer's door. A corporation might own a lab to invent products, a factory that made parts for products, another factory that assembled the parts into a whole, a distribution system, and maybe the retail stores. It meant building huge scale, which took time and a lot of money. Once erected, these big-scale barriers to entry made it hard for newcomers to com-pete because it was supremely difficult to build all that scale.

By the 1990s, with the arrival of the widespread use of computers, the internet, and globalization, we began to see cracks in the founda-tions of vertically integrated corporations—the first intimations of un-scaling. Companies discovered they could outsource entire functions and whole departments to other companies and even other countries—the connected outsourcing dynamic behind the sentiment that "the world is flat," as author and *New York Times* columnist Thomas Fried-man put it. The more platforms we built using new technologies, the more companies could rely on those platforms to do a job or task in-stead of doing it themselves. Barriers to entry kept falling. New en-trants could be smaller and instead use platforms to seem big. Consider how upstarts like online eyeglass company Warby Parker or Jessica Alba's consumer health and wellness goods company, The Honest Company, were able to quickly use the internet to sell to a global mar-ket to compete against established eyeglass makers and consumer prod-ucts giants. The new era of startup-driven disruption took shape.

Around 2007 the creation of platforms accelerated. Smartphones and mobile networks allowed new services and products to reach almost any-one, anywhere. Social networks exploded and gave companies new ways to find people and advertise to them. Cloud computing meant any com-pany could start a computing-intensive digital company without ever buying more than a laptop—just click a few settings on Amazon Web Services, enter a credit card number, and start selling to the world. At the same time, more businesses became digital—music, news, online retail, software as a service. Digital businesses especially could utilize platforms to instantly create, make, market, and deliver products anywhere in the world. As more business became digital, companies could collect more data about almost everything—customers, products, transactions, logistics—and that data made software and platforms smarter, creating an accelerating positive cycle. As this trend sped up—building more digital platforms, turning more business into digital business, and generating more data—we hit an inflection point. We started to reinvent the dynamics of business.

By 2017, ten years after the iPhone, platforms could do almost *everything* a business might need. One person could start a global company in her basement and compete against giants just by renting everything that major corporations used to need to build for themselves. Warby could rent computing power on a cloud service, rent ways to reach consumers via social networks and search engines, rent production from contract manufacturers, rent distribution of its glasses through FedEx and UPS, and so on. This is the essence of unscaling: *Companies can rent scale. They no longer need to own it.* And that changes everything.

Unscaling, it is important to note, is only beginning. As AI and other new technologies emerge and get developed into platforms, tiny entrepreneurial companies that have yet to be founded can serve customers in ways that big, mass-market companies could never imagine. Entrepreneurs will increasingly plug into platforms to build superfocused products that greatly appeal to niche markets, then find passionate customers and sell to them anywhere in the world—and do it all at profit margins that once only came with the old economies of scale. Big companies, bogged down by their own scale, will find it increasingly challenging to win against highly specialized, fast-changing products and services. That's why the forces of AI and unscale are taking the twentieth-century economy apart and reassembling it in an entirely different way.

* * *

The emergence of the AI engine underneath unscaling is a grand technology story. In 2007 Apple introduced the iPhone. There had been smartphones before with brands like Blackberry and Nokia, but they didn't have anything close to the iPhone's capabilities. More importantly, Apple introduced the concept of the *app*. Over the following decade the mobile device moved from being an accessory to becoming the main way most people use software, data, and connected services—which, significantly, were hosted in the cloud. Before 2007—heck, even in 2010—*cloud computing* was a nerdy tech concept most people didn't comprehend. Now most people know it as a handy reference for the fact that most of our data and the software we use sits on some computer in a gigantic data center somewhere, and we connect to it through wireless networks.

A number of other important technology platforms emerged around 2007 and took hold in the years after. When Amazon.com, which had already moved commerce online, launched Amazon Web Services (AWS) in 2006 it gave every software developer the power to launch a cloud-based software product and become an entrepreneur. Facebook was founded in 2004, but it wasn't until 2007 that it turned into a platform, opening up so developers could build applications on it. Added together, 2007 can be called the origin point of an AI revolution, made possible by the combination of mobile computing, cloud computing, and social networking. In 2007 a little more than 1 billion people were on the internet; by 2016 it was 3 billion. Smartphone use had grown from a tiny sliver of society in 2007 to more than 2.5 billion people in 2016.

The new platforms made it possible for a new generation of entrepreneurs to reimagine how we do things and to build disruptive new apps. At first the platforms gobbled up cameras, flashlights, maps, publishing, music—all now on your phone or in the cloud, generating data. Because of the smartphone and cloud, Travis Kalanick and Garrett Camp could turn their frustration over waiting for a taxi in Paris into something productive by reimagining ride hailing through an app—giving birth to Uber. The concept of a cloud-based social graph gave the founders of Airbnb a way to connect people with places and to build in a system of trust. Two young brothers from Ireland, John and Patrick Collison, saw a way to use the cloud to offer developers a way to take payments anywhere in the world and founded Stripe. Evan

Spiegel could reimagine communications as something more ethereal than it had been on the internet and started Snapchat.

In ten years a few important technology platforms completely transformed the way 3 billion people work and live. As content, community, and commerce continue to move online, we're collecting data we never had before—data about what you buy, what you read, who you know, where you go. That data gives companies exciting new insights that can lead to even more new products and services. And it feeds the machine learning of AI software, which constantly gets better the more it is used because every interaction teaches it more about whatever the software is programmed to do.

* * *

I didn't fully understand what was happening in 2007 and nearly missed it. Let me explain why—and how I came back around to understand the force of unscaling. The story goes back to New Delhi, where I grew up.

My parents were smart enough to recognize that there was no level playing field for a family of our means in India. They didn't have the resources to provide a world-class education for me and my sister. So when my uncle sponsored us for a green card in the United States, my parents gambled everything to give my sister and me the opportunity to thrive in a more egalitarian society. This was the biggest risk we ever took as a family—and it probably helped me understand the value of taking risks based on a vision of how things can be, which is really what I do as a venture capitalist (VC).

Our early years in America weren't easy. We had to live in the basement of a home in Brookline, Massachusetts. Sleeping next to a boiler wasn't exactly fun! Moreover, I had to get a job at a local CVS during high school and work extremely long hours to help meet our financial needs. But none of it felt like a hardship because I felt really energized about my new school. Coming from India, I was amazed to have the opportunity to pick which courses I wanted to take! In India we had no choice of classes. In the United States I went overboard and, in a self-directed and self-paced fashion, completed science and mathematics requirements for my freshman year of college before finishing high school. My experiences stuck with me and inform how I think about personalized education in an AI-driven, unscaled era.

I went to Massachusetts Institute of Technology (MIT) and continued with my self-paced, self-directed mindset. I decided to learn at my own pace and take classes from as many departments as I could. I remember recognizing early on that being a straight-A student, although a tremendous accomplishment, wasn't going to matter in the long run. So I would regularly skip classes, often joking with my friend Sal Khan that classes were always too fast or too slow for me. Well, at least that was my excuse for skipping the classes. Years later Sal went on to start Khan Academy, with self-paced learning as his early leverage point for transforming education.

By senior year I had finished an unusually large number of courses but didn't have enough credits to meet the requirements for any individual department. (Eventually I wound up with degrees from MIT in six different disciplines.) I wanted to know different disciplines and be able to think across them. In my career this method of "systems thinking" has helped me connect disparate signals from various parts of the economy to see larger trends.

At the turn of this century the development of mobile web technologies inspired me, leading me to drop out of my doctoral program at MIT. My mother still hasn't forgiven me! I became an entrepreneur in the mobile space and, along with some friends, started a software company that built tools to simplify how mobile applications are developed. Our mission was to help developers rewrite software originally written in the 1980s and '90s to make it more natural for people to use.

In retrospect we gave up on our mission too early, selling the business after a couple of years. The handsets and communications networks of that time weren't ready to power such a vision—though they would've been a half dozen years later. After selling the business in 2001 I joined General Catalyst, then a year-old venture capital firm in the Boston area. I started investing in traditional software, but the companies just weren't that impactful. So I searched for a new, grander frontier, and I decided on energy, aiming to solve climate change. This is where I whiffed on seeing the mobile-social-cloud revolution and the promise of AI: I went into energy around 2006 in part because I thought software technology had stalled. But investing in the energy industry also taught me important lessons that led to how I now think about unscaling. In a regulated industry companies are too often motivated to serve the regulators, not the customers, and the economics of a regulated company give the company little incentive to innovate. That can make regulated sectors like energy a ripe target for entrepreneurs with fresh ideas.

All these experiences increasingly came together in my work investing in tech companies, and I understood there is a pattern: each industry I've become involved in is not going through its own unique transformation; rather, the entire global economy is going through a transformation and dragging all these industries and sectors with it. We're shifting from mass-produced products that can be sold to the most people possible to highly personalized products that delight small niches of passionate customers—at prices that are often lower than the mass-market products. And of course customers will choose personalized products over mass-market products—because personalized products are by definition geared especially for each customer. The unscale mindset asks: What can I build that makes each individual happy? That's a big change from last century's mindset of: What can I build to sell to the most people?

When I realized this shift was happening, I moved to Silicon Valley to be closer to the entrepreneurs driving unscaling. My first investment was in payments company Stripe, whose founders moved to the Bay Area from Boston just around the same time I did. Stripe was focused on helping new online businesses process payments with great ease, giving small companies anywhere on the planet the payment processing benefits that usually come from a big bank, yet built in software and much less costly than the fees banks charge. Since then various companies have emerged that, collectively, can run the back office of a new startup with the same sophistication as a *Fortune* 500 company. These kinds of platforms allow startups to focus on what matters: *delighting customers with great products and services.*

My view of the future really shifted after one memorable meeting in 2012. Our firm had hired a fresh graduate from Stanford who told us about a couple of students still on campus building this interesting app that allows users to send texts and photos that would then disappear—an app later named Snapchat. We set up a meeting with the students, Evan Spiegel and Bobby Murphy, and together we riffed on what their idea meant in the grand scheme of things. Evan made me realize that for almost two decades we'd been having digital conversations that were extremely unnatural. For most of human history, when we talked to each other, the conversation left no record. It couldn't be copied and sent to others or analyzed for advertising—in other words, it wasn't like email, Facebook posts, chats, or tweets. Snapchat would give us a way to have electronic communication that would be more like face-to-face conversation, leaving no record, no trail.

That's when the first big thought hit me: we're entering an era when technology could finally conform to humans rather than the other way around. From there I rewound and rethought what I've learned about technology and realized that we are about to recreate just about everything. From 2007 to 2017—thanks to mobile, social, and cloud—we made computing and connectivity nearly ubiquitous and infinite around the world. Computing power is basically on tap through the cloud—you can get all the computing you need. Connectivity in much of the world is a given—cheap, easy, and available almost everywhere.

* * *

Moore's Law has long described the speed of change in computing. Gordon Moore, a Silicon Valley pioneer and cofounder of Intel, explained in 1965 that the number of transistors in a microprocessor can double every eighteen months for the same price of product—which has meant that the power of computing can double every eighteen months for the same price. That dynamic made computers relentlessly better and cheaper and drove them into everyday life.

Then, in the 1980s, Bob Metcalfe, who is credited with coinventing Ethernet, one of the earliest computer networking systems, described the exponential power of networks by showing that the value of a network is proportional to the square of the number of users connected to it. That exponential dynamic meant that as more than 3 billion people connected to the internet from 1995 to 2015, the internet exploded in power and value—creating a societal and economic impact far greater than just the number of people connected. Moore's Law made computing affordable and accessible, so now every person or thing can have computing power. Metcalfe's Law made it valuable to move content, community, and commerce online. Those technologies have been unbelievably powerful forces driving change. But now Moore's Law and Metcalfe's Law are reaching diminishing returns. The laws of physics mean that microprocessors can't get much smaller and faster anymore, and if most of the world that will ever get connected is already connected, the benefits of Metcalfe's Law taper off.

But there's a new post-Moore/Metcalfe dynamic kicking in. The cloud is essentially the meeting point of Moore's Law and Metcalfe's Law—where data, computing resources, and connectivity have merged. It is now affordable to put a microprocessor into everyone's pocket and inside everything—and because almost every person and everything is connected to the internet, we can have a real-time feedback loop with them all. This feedback loop powers unscaling because it allows AI-driven software to continually learn about customers and the world so companies can deliver exactly what individual customers want.

At the intersection of Moore's Law and Metcalfe's Law, unscaling of the economy is proportional to the connections into the cloud. The economy's ability to unscale grows linearly with every new connection into the cloud. These trend lines are coming together to drive a new technological age.

We went through a similarly grand revolution in the early twentieth century when the telegraph, telephone, television, automobile, airplane, and mass electrification vaulted the planet from a slow and local way of life to one that is fast and global. No one who lived in the late 1800s would have recognized the world of the 1920s.

Economist Carlota Perez describes the impact of such revolutions in her influential book *Technological Revolutions and Financial Capital*: "When a technological revolution irrupts in the scene, it does not just add some dynamic new industries to the previous production structure. It provides the means for modernizing all the existing industries and activities." In her construct, today we're in the *installation* phase of AI-driven unscaling, "during which the critical mass of the industries and infrastructures of the revolution are put in place against the resistance of the established paradigm." Over the next two decades this revolution will hit a *turning point* and then shift into *deployment*, "leading ultimately to a different 'way of life,'" as Perez states.

As I write this, many companies—including IBM, Google, Facebook, Amazon, and Apple—are racing to create AI platforms. Similar races are on to develop virtual reality and augmented reality platforms. The same could be said of the Internet of Things (IoT), genomics, blockchain, and 3D printing. All these technologies—and many more—are likely to turn out to be even more important than the 2007 development of mobile, social, and cloud, and they will build on each other in a compounding effect.

Every kind of industry will be affected, even those that seem ancient and impervious to digital transformation, such as healthcare and energy—and even government. If the forces of unscaling were first set loose from 2007 to 2017, they will become ten times greater from 2017 to 2027 because of the compounding effect of the technologies we are creating.

* * *

Although it's impossible to predict all the outcomes of unscaling, there are aspects of an unscaled world we can anticipate. Put together mobile, cloud, Internet of Things (IoT), augmented reality, software, and AI, and we wind up with a completely connected and instrumented planet—essentially creating one global system of people, places, and things. We will be able to get data about almost anything and understand far more about how the world works on both a macro- and a microlevel.

Unscaling will involve transitioning away from ownership and toward accessing services. So, for instance, transportation will become an on-demand utility. Owning a vehicle is an expensive part of most people's budget. If you live in an urban environment, chances are, you won't need to own a car. You'll probably use Uber-like services, and a self-driving car will pick you up. People who own robot cars will let their cars work for Uber (or its successor companies) during the 90 percent of time a private car otherwise stays parked. It seems highly likely that within twenty years the number of cars on the road and in parking lots will decrease, and traffic deaths—now thirty thousand a year in the United States—will fall precipitously.

The key to success for most people will be living an entrepreneurial life and becoming their own personal enterprises, selling services on demand through the cloud to many employers. That's not just for business owners but for everyone. For better or worse, a decreasing percentage of the population will rely on traditional full-time employment—and an increasing percentage will do better by owning their own business, with overlapping mini-careers throughout their lives.

More of the global population will access education through on-demand services—whether for K-12, college, or lifetime learning—available on any device or, eventually, in virtual reality. It's already happening. AI-guided courses like those from Khan Academy and Coursera already supplement college educations and help people learn throughout their lives. Before long, unscaled learning will begin to disrupt our highly scaled system of big colleges and big high schools. A lot of people struggling today to pay off tens of thousands of dollars in college debt might already wonder about the value of four years on campus.

Healthcare is on its way to becoming more preemptive instead of, as it is today, reactive. Newborns will routinely have their genome sequenced, and that data will help predict diseases. IoT devices will be able to monitor your vital signs and activity, spotting problems at a very early stage. You'll be able to get an initial diagnosis from an AI software "doctor" app via your phone or some other device, and the AI will guide you to a specialist if needed. Healthcare will be flipped on its head, shifting from treating health problems after they arise to spotting and fixing them before they develop. That should cost a fraction of what healthcare costs today, solving one of America's toughest financial squeezes.

As entrepreneurs remake the energy sector, more homes and buildings will generate their own power using cheap and superefficient solar panels on roofs and high-powered batteries in basements or garages. The batteries, like those now being manufactured by Tesla, will store power generated when the sun shines for use when it doesn't. Each of these buildings will be connected to a two-way power line that can allow anyone to sell excess energy or buy needed energy in an eBay-style marketplace. If you do own a car, it's likely to be electric, and your home solar panels and batteries can charge it.

Trends suggest that you will get more of your food from small local farms or urban farms built inside old warehouses and shopping malls. The food industry spent the twentieth century scaling up agriculture, making farms bigger and more corporate, tended by enormous pieces of machinery with very few actual farmers. In the coming decades technology will help small, local farms operate at a profit, while breakthroughs in producing test-tube meat will vastly reduce the acreage needed to graze cows and raise chickens.

The technology of 3D printing is beginning to unscale and reimagine manufacturing. Within a decade, if you order a new pair of shoes or a chair, it might not come from a far-off mass-production factory; instead, many companies are going to custom produce items in small batches as they're ordered, and the factories will operate akin to AWS— offering companies as much or as little manufacturing as they need.

Unscaling will leave few industries or activities untouched. Whatever kind of work you do or wherever you live, your journey will be different from that of past generations because of AI and unscaling. In the rest of this book I'll go into detail about the many ways it will be different and how to think about those changes with an unscale mindset so you can take advantage of what's coming.

* * *

We have a lot of choices to make as we decide to shape the future. Unscaling is disruptive. It is remaking an old economy into a new one. Whenever that has happened in history, whole categories of jobs disappear, and it will be no different this time as AI automates many new tasks. Donald Trump was elected president of the United States in 2016 largely on a wave of unhappiness over jobs and economic disruption. And the anxiety is only going to intensify. A 2016 Pew Research survey found that one in five of those people with a high school diploma or less believes they're in danger of being replaced by AI software. A research paper from Oxford University proclaimed that machines will take over nearly half of all work currently done by humans. The media has been packed with stories about AI eventually leaving people with no work to do. So AI and unscaling are going to force us to rethink what it means to work and make a living. It might force nations to consider instituting guaranteed incomes or to make sure education is free and easy to get instead of—as it is today in the United States—expensive and increasingly available to only the top levels of society. As the Trump election showed, if technologists and policymakers don't deal with these issues and help pull people through this disruption, the people left behind will rebel and try to stop or reverse unscaling.

As AI software runs more of our lives, algorithms need to be held accountable in the same way we hold people accountable, preventing automated discrimination or even criminal acts. The algorithm behind Facebook's news feed is optimized for making money for Facebook, not for ensuring fairness or civility; this has arguably led to a greater political divide during this last election cycle in the United States. That's just one early example of how algorithms with no moral guidance can impact our society. We'll need to decide whether we want our businesses to make their algorithms more publicly accountable. Remember, companies optimize their AI software for profit to serve their shareholders, not for doing the right thing or making decisions transparent. This needs to change, and these companies themselves must take the lead in creating algorithmic accountability in their services.

Several big projects in longevity are aimed at extending life expectancy by decades. Google's Calico is putting $1.5 billion into discovering the basic science behind aging, the Jeff Bezos–backed Unity Biotechnology is investigating drugs to rejuvenate aged tissues, and we at General Catalyst invested in Elysium Health, a company with a stable of expert aging and bio-scientists focused on boosting cellular NAD+, a critical coenzyme that begins to decline in our twenties. If we use AI to automate vast numbers of jobs and stick the landing on longevity, then what happens? Will we be asking formerly employed drivers to while away decade after newly found decade with no potential for work? My friends Sam Altman, who runs the tech incubator Y Combinator, and Chris Hughes, a cofounder of Facebook, have kicked off two separate universal basic income (UBI) projects that explore replacing employment-derived income with unconditional stipends. Both are trying to get ahead of the impacts of the highly automated, postwork world we're headed toward. But although UBI replaces monetary loss, it does not address something just as fundamental: *purpose*. Exciting as it is to be turning science fiction into reality, once most of the labor market is automated, as humans we will need to find a fulfilling way to spend our 120-year lifespan.

Another concern in the new tech age is monopoly power. In digital industries, more than in physical-product industries, the tendency is toward winner-take-all. That could lead to monopolies controlling vital parts of the economy—as we've seen Facebook dominate social networking and Google dominate search. If we're not careful, such monopolies could impose rules and practices that benefit them to the detriment of society.

Not to be alarmist, but in the early 1900s—the last time technology so completely transformed the economy and life—the shocks were followed by two world wars, a global economic depression, and the rise of a Western-led liberal world order. The changes being wrought today are even more dramatic. We need to hope our leaders can avoid war, but turmoil will accompany us through this journey as some voters and governments struggle against change while others embrace it. In the last century aggressive countries fought wars over natural resources, especially oil. Perhaps the next wars are going to be fought over data, and the rise of global hacking is a precursor.

Given the right choices, however, I believe the AI century can be enormously beneficial. AI-driven unscaling will be all about creating products and services that are better, cheaper, and easier to get than ever before—tailored specifically to you.

If we make good choices, we should see an improvement in quality of life. Most technology to date has been about efficiency—encoding a task in software to automate it. As industries unscale, software will move to the next level and make products and services more effective, in the way Uber has made getting a ride not just more efficient but simply *better* than hailing a taxi. Imagine the most frustrating aspect of your life, and now imagine it getting better, cheaper, and easier to get.

How we train our students for this world will be critical to securing their future employment. They need to find the next thing humans can do that machines can't—no doubt involving "human" capabilities like creativity and psychology—and learn to collaborate with AI-driven machines in ways that unleash human potential.

The bottom line: we have choices to make about where to point innovations, how the workforce will evolve, and how we make sure the algorithms uphold our values. There has never been a better time with more opportunities and lower barriers for enterprising people and organizations. We're at the cusp of an amazing adventure. We have a chance to rewrite our world and solve some of the greatest problems we face, from climate change to cancer. As with the last technological revolution, by the time we're finished, the planet will be almost unrecognizable.

2

The AI Century

Artificial intelligence is this century's electricity.

As the twentieth century was just dawning, radical new technology, much of it driven by electricity and oil, cascaded into people's lives, changing society in ways no one could have anticipated. That flurry of technology set in motion one hundred years of scaling up.

In the 1880s small electric stations built on Thomas Edison's designs were spreading to cities, but each could only power a few blocks of buildings. At the end of the 1890s New York patent attorney Charles Curtis developed the steam turbine generator, which for the first time allowed mass-market electricity to be produced inexpensively. In the early 1900s electric grids started to crisscross cities in a building spree.

The spread of electricity allowed factories to set up anywhere, altering the way manufacturing centers formed. Lights allowed factories to operate at night. And electric power made the modern assembly line possible. With electricity Guglielmo Marconi completed the first two-way wireless message—a fifty-four-word greeting to England's King Edward VII. Electricity powered telephone exchanges. Alexander Graham Bell invented the telephone in 1876, and it took hold in cities in the 1900s. Communication allowed companies to scale up more because they could now better coordinate more people over longer distances.

Other bold technologies arose. From 1900 to 1902 the Germans invented the zeppelin airship, George Eastman developed the first consumer camera, salesman King Gillette created the first safety razor, and the first electric stoves made their way into homes. People could do things they'd never done before—fly, travel, take pictures, light their homes with electricity—and they wanted more.

Into this milieu strode Henry Ford. In the late 1800s he had worked at the Edison Illuminating Company, where he met and was inspired by Thomas Edison. By night he experimented with motorized quadricycles. He started two automobile companies in three years, and both failed. In 1903, just before his fortieth birthday, he founded Ford Motor and began making the Model A. The effect on popular thinking is hard to imagine. In 1903 horses were so prevalent that every day in New York 2.5 million pounds of horse manure was deposited on the city's streets. In 1908 Ford unveiled the Model T, which blew the lid off the industry, albeit with a modest start of 239 cars sold that year. The number increased in 1909 to 12,176. By 1910 one cartoonist's depiction of the future showed grade-school students driving tiny cars to class. By 1913 Ford's sales mushroomed to 179,199 cars, and the numbers shot up from there. And it was all about scale. The Model T famously came in one color: black. It was the first mass-produced car for a homogenous mass market.

In Dayton, Ohio, the Wright brothers built on advances in engine technology and new ideas about winged flight. In their bicycle shop they and their mechanic, Charlie Taylor, built an engine and married it to a flyer with a wingspan of about forty feet to be the first, in 1903, to successfully fly. By the 1930s Pan Am began circling the globe with civilian air travel.

Everyday life was being completely transformed. Big electricity-driven factories made mass-market products to fill the shelves of giant department stores like Macy's and Sears. Radio opened up the concept of mass-market advertising. To build massive amounts of physical products, ship them, sell them, and advertise them through a limited number of media outlets, companies needed to get big—and once they were big, their scale became a barrier to entry for newcomers.

The technologies of the early 1900s created a mass-market consumer culture and began a century of building economies of scale. The corporation became the centerpiece of global business, with the goal to get as big as possible. The *Fortune* 500 list—an unabashed celebration of scale—debuted in 1955. General Motors topped the list, and it had 576,667 employees. In 2016 Walmart was number one on the list—with 2.3 million employees. Governments scaled up too. The US federal government employed about 1 million people in 1900 and more than 4 million in 2015. In Hollywood, movies had to be blockbusters or bust. Massive brands like Budweiser, Coca-Cola, and McDonald's served the same thing to everybody and wiped out niche competitors, while Walmart nuked local retailers by building ever-bigger stores in ever-more places. The Western world scaled up mass-education schools, modeling them on assembly lines—children entered kindergarten and would move through the system one step at a time, all learning mostly the same things, until they popped out the other end, finished and ready to work.

Scale helped society accomplish great things—educating the masses, improving quality of life, eradicating diseases like smallpox, lifting millions out of poverty. Twentieth-century technology, enabled by electric power, made it all not just possible but inevitable.

* * *

In 2007 artificial intelligence had already been around for decades, and at various times—the 1980s, the early 2000s—AI was supposedly about to break through and change the world. But it didn't. The technology simply didn't have enough data to learn from, so it couldn't get past some specialized uses (like the autopilot on airliners). But around 2007, as we started to move more of our lives onto mobile, social, and cloud platforms that could collect enormous amounts of data, AI could finally make an impact that rivals that of electricity in the early 1900s. Much the way we electrified the world to set the previous scaled era in motion, we are infusing AI into the world today, and that has set in motion unscaling.

Hunch, a business founded by Chris Dixon, marked the start of my involvement in AI-related companies. As an undergraduate at Columbia in the early 1990s, Dixon majored in philosophy. He went on to get his MBA and became a software developer for Arbitrade, a hedge fund that focused on high-frequency trading. Dixon then started a company called SiteAdvisor, which helped internet users avoid unwanted spam. SiteAdvisor was the first investment I ever made as a venture capitalist. In 2006 we sold the company to security software firm McAfee.

The next year, 2007, Dixon and two cofounders started Hunch, and I invested. We didn't call it AI or machine learning or cognitive computing, but Hunch was a web application built on software meant to learn. Today we would certainly call it AI. The goal at Hunch was to build a "taste graph" of the internet, linking people to things they liked, whether it was a product or a singer or a website. If we built a big enough taste graph, Hunch could learn from everyone's likes to make accurate recommendations to its users. It could figure out that if you liked Beyoncé and chili dogs and Southwest Airlines, you might like, say, clothes from Urban Outfitters—because other people who have tastes similar to yours also like Urban Outfitters.

The technology worked well. But we ran into a challenge: AI needs to learn from enormous amounts of data, so the more data, the better the AI. As an independent entity, Hunch just could not get enough data from enough users to make the AI sufficiently effective to convince even more users to join Hunch—which might've kept making the AI better in a virtuous cycle.

In 2011 an excellent path forward surfaced. We sold the company to eBay for $80 million. At the time, eBay had 97 million users, 200 million active listings, 2 billion daily page views, and around nine petabytes of data about all of that activity. "With eBay's data behind us, expect Hunch to get much, much better," Dixon said when announcing the deal. We finally had the data to train our AI to be really good. "Hunch discovered that a certain class of users who were buying gold coins were also the perfect customers for a microscope that they could use to examine those goods," eBay chief technology officer Mark Carges told the press after the deal. "That is the kind of odd association we never would have found on our own." Importantly, this taste graph is exactly what Facebook implemented with its "like" button—and did so much more effectively because it assembled more than a billion users who clicked "like" buttons constantly throughout each day.

Hunch's journey was very much a sign of the times. Although the concept of AI had been around for six decades, it didn't have the data to make it really good. Now we have it, thanks to the technologies we built over the past decade, starting around 2007.

Think about how, in less than ten years, life changed dramatically. By 2016 more than half the planet owned a smartphone, and mobile networks emerged as an awesome new platform. On a phone you can download apps that connect through a nearly ubiquitous high-speed wireless network to powerful software hosted in a data center somewhere. Those apps deliver social networking, chat, email, shopping, media, and services such as Uber or Airbnb. Super-accurate GPS maps guide you anywhere. The phone can hold your music and books. It can let you watch live sports. Enterprise apps from companies such as Salesforce.com allow you to do your job from a phone screen almost anywhere in the world. Mobile content, community, and commerce have quickly become a way of life. We can't imagine getting through the day without this technology. How strange to think that just a decade before, none of this was possible.

So we spent a decade connecting people and moving a great deal of our activities online, where every action generates data. Then, in recent years, industry jacked up the planetwide data-generation machine by implementing the Internet of Things (IoT), which puts "things," not just people, on the global network. Those things might be video cameras or heat sensors, Fitbit health monitors or GPS tags on endangered animals. According to analyst firm IDC, there were already 9 billion connected devices in place in 2015, but that will grow to 30 billion by 2020 and 80 billion by 2025. Technology giants such as Cisco, IBM, and General Electric are investing enormous amounts in IoT sensors and data. I invested, alongside industry legend Marc Andreessen of VC firm Andreessen Horowitz, in a startup called Samsara that's building a next-generation IoT platform to collect and manage data from sensors. As sensors get embedded into almost everything, the technology will create a kind of quantified planet. In the 2000s, as people in Silicon Valley liked to say, software was eating the world, moving into every nook and cranny of business and life. But in the 2010s and beyond, the world is eating software—everything is ingesting software, becoming smart and connecting to the global internet.

One way to get a sense of the impact of IoT on data collection is through the common light socket. One estimate says there are 4 billion street lamps in the world. There are another 4 billion household light sockets in the United States alone, an average of fifty-two per house. Add in businesses, schools, airport terminals, and so on, and we have tens of billions of sockets spread across every populated area. Each socket is a source of electricity, which can power sensors and wireless networking devices embedded in a bulb. Lights were once dumb and isolated, but they're now becoming smart and connected, collecting truckloads of data about everyday life that can help AI get smarter. IoT devices are spreading everywhere—into jewelry and clothes, sewer pipes and waterways, even tags on animals in the wild and pets at home. In industrial settings sensors throughout factories or in assets like trucks or jet engines can all feedback through a system like GE's Predix, a cloud platform that connects things and people. IoT is invading buildings: Japanese manufacturer KONE is adding IoT to its elevators, escalators, turnstiles, and automatic building doors that move a billion people each day. Kimberly-Clark is developing the Scott Intelligent Restroom. The system is a full-building network of washroom product dispensers that monitors almost every aspect of the system status, transforming the way washrooms are maintained and supplied. If you're in a Scott-monitored building, you will never again wash your hands and find no towels in the dispenser.

The IoT explosion will give us an astonishing flow of data, opening possibilities for AI to conduct deep analysis of how the world works. IoT is giving us instant, real-time views, as if we are hooking up the planet to an EKG and watching its heart beat.

AI is all about discerning patterns in data, predicting behaviors, and deciding on actions. With little data coming in, AI is like a baby's brain—all the smarts are in place, but it has too little knowledge of the world to understand what's going on or to know that, say, pulling the cat's tail will get you scratched. So, like a brain, as AI gets exposed to greater amounts of data, it can see patterns and predict behaviors with greater accuracy. The more data coming in, the better AI gets.

We already encounter AI all the time. Google's AI-driven search algorithm learns from every search and gets better. Facebook's AI learns from your posts and likes and then populates your timeline with feeds and ads you'll probably want to see. Netflix learns from your viewing habits, matches it with what it's learned from millions of other viewers, makes recommendations, then uses that AI-based learning to guide its decisions about what movies or series to produce in order to appeal to the most users. Hedge funds rely on AI to see trading patterns humans could never find. Security software relies on AI to learn about normal activity in a system so it can recognize and stop an intruder. In 2016 IBM bought The Weather Company, which gathers incredible amounts of weather data from sensors all over the planet, so IBM can feed its data to IBM's Watson AI. Now Watson can literally "learn" how weather works and make hyper-local microforecasts—for instance, predicting the wind patterns at the location of an outdoor Olympic diving event.

As I write this in 2017 Google, Tesla, General Motors, and others are developing self-driving cars. AI makes the technology possible, and the more cars run on auto-pilot, the more data those AI systems will collect, making the autonomous vehicles ever better. Personal assistant bots like Amazon's Alexa and Apple's Siri are making their way into everyday life. Those AI programs now have so much data from listening to people speak that their ability to recognize spoken words is better than a human's. They still have trouble understanding complex questions or commands, but the more people use them, the better they will get. That's the nature of AI.

AI is foundational in almost all the entities I fund or work on. Khan Academy is using AI to change the way we learn so coursework can tailor itself to the student's pace of learning. Livongo, as described earlier, is building AI into its software to learn about the health and actions of people with diabetes so it can better help each individual manage their condition. I can't imagine investing in something that doesn't use AI.

So much of what we do now generates data because we do it online, with the help of a smartphone, or by touching something embedded with connected IoT sensors. Data is the raw fuel of this new revolution, just as fossil fuels and electricity were needed for the 1900s Industrial Revolution. AI takes data and makes it useful and accessible—the way electricity became useful power that could flow into almost anything, anywhere. AI is getting built into everything that has computing power. In another ten years anything that AI doesn't power will seem lifeless and outmoded. It will be like an icebox after electric-powered refrigerators were invented.

By the mid-2010s investment was pouring into AI. Funding for AI startups soared to more than $1 billion in 2016, according to analyst firm CB Insights. That was up from $681 million just the year before, $145 million in 2011—and just a trickle before that. (You would have had a hard time finding a company that described itself as an AI startup back when I invested in Hunch.)

Society has come to *need* AI. The world's systems have gotten so complex and the flood of data so intense that the only way to handle it all will be to employ AI. If you could turn off every AI program in use today, the developed world would shut down. Networks would seize up, planes couldn't fly, Google would freeze, spam would overrun your inbox, the Postal Service couldn't sort mail, and on and on. As the years go by, AI will become yet more integral to keeping the planet's systems running.

The best AI software will evolve into our trusted collaborators. AI software in a conference room will be able to listen to the conversation in a business meeting, constantly searching the internet for information that might be relevant and serving it up when asked. "It can bring in knowledge of the outside world that the humans might not be aware of," and that means the humans can make better decisions, says Surya Ganguli, who researches AI and brain science at Stanford University's Neural Dynamics and Computation Lab. "That's a huge frontier. It could know what was said in the room an hour ago, know the history of the field, the goals of the people trying to solve the problem, and figure out a suggestion."

By the early 2020s AI will be better than healthcare providers at diagnosing medical problems and better than legal assistants at researching case law, Ganguli tells me. "Artificial intelligence is more than legal technology," says the American Bar Association. "It is the next great hope that will revolutionize the legal profession." IBM's Watson is already becoming a doctor's assistant, ingesting libraries full of medical research to help diagnose patients.

Scientists all over the world are working on mapping and understanding the brain. That knowledge is informing computer science, and the tech world is slowly creeping toward making computers that function more like brains. These machines will never need to be programmed. Like babies, they will be blank slates that observe and learn. But they will have the advantages of computers' speed and storage capacity. Instead of reading one book at a time, such a system could copy and paste every known book into its memory.

Jeff Hawkins, the CEO of brain-like software company Numenta (and the guy who invented the PalmPilot) explains, "We have made excellent progress on the science and see a clear path to creating intelligent machines, including ones that are faster and more capable in many ways than humans." As an example, Hawkins says we can eventually make machines that are great mathematicians. "Mathematicians try to figure out proofs and mathematical structure and see elegance in high-dimensional spaces in their heads," he says. "You can build an intelligent machine that is designed for that. It actually lives in a mathematical space, and its native behaviors are mathematical behaviors. And it can run a million times faster than a human and never get tired. It can be designed to be a brilliant mathematician."

AI is already starting to get built into platforms. IBM lets companies big and small develop products built on top of Watson. Amazon's Web Services will increasingly have AI capabilities built in, as will cloud computing services from Google and Microsoft. That will allow any startup to rent AI capabilities using nothing but a credit card and then build AI into almost any app or service. And as AI improves, it will drive more unscaling. AI allows the profitable customization of everything—because AI can automate customization. Think about it: people have always had products and services tailored just to them, but if humans need to do the tailoring, it takes a great deal of time and labor. So custom-made products or a personalized service like having your own driver can only be offered profitably if they cost a lot—too much for mass-market consumers to afford. Mechanized factories pay off because they can make a lot of the exact same item quickly and cheaply.

But AI is different. AI automates learning about each individual customer or user, and then software can automatically tailor a product or service to them. An AI-based service from Livongo can understand how to treat your diabetes on an intimate level. AI-based Uber can offer an on-demand car ride whenever you want, just like a personal driver, but at a fraction of the cost—and eventually Uber's cars will be autonomous AI-guided robots. AI creates the conditions for the opposite of a mass market. It is flowing into every kind of technology, product, and service and makes it possible to profitably serve a market of one. And so AI creates conditions for the opposite of scaling up. A product tailored to you will beat a product made for the masses. Scaling up was the way to economically produce mass-market products, but an AI platform added to mobile, social, cloud, and other twenty-first-century platforms makes it possible for small, entrepreneurial companies to quickly and simply develop, sell, and deliver market-of-one products. Scale no longer gives companies an automatic advantage when unscaled and focused companies can tap into platforms and build products and services for a specific audience.

* * *

AI is powering all sorts of profound new technologies and companies that will change work and life.

In the mid-2010s virtual reality (VR) and its cousin, augmented reality (AR), graduated from geek dream to a viable technology. My aha moment came when Oculus Rift, maker of the first VR goggles, was trying to raise money on Kickstarter to build its first product. I got a demo of the prototype and realized that we are going to create a virtual online world parallel to the real physical world. As people spend more time in virtual worlds, more demand will be generated for services, art, games, and entertainment inside those worlds. The economies of unscale will be even more pronounced in a virtual world because absolutely everything in a VR world will be digital and every action will generate data, which, in turn, improves any AI-based product or service. As I thought about how to invest in VR and AR I realized that companies and individuals are going to want tools to help them create anything, from virtual buildings and furniture to services inside virtual worlds, whether for fun or for profit. And that, in turn, led me to Angle Technologies.

While students at Harvard in the mid-2000s, David Kosslyn and Ian Thompson hung out together, often hacking code, sometimes playing Minecraft. Thompson in particular followed developments in virtual reality. Thompson remembers when an architect friend had him put on goggles and virtually tour a mock-up of a new Bay Area Rapid Transit (BART) train station, the mass-transit system that serves the San Francisco Bay Area. "He dropped me in, and it was nauseating"—a common problem with VR in its early years—"but it was also amazing! I got hooked," Thompson says. Thompson shared his passion with Kosslyn, and the two tossed around ideas about what they might create in VR.

Kosslyn went on to work at Google and YouTube, while Thompson bounced through a few startups. But they kept talking VR and watching it improve. And then, in mid-2014, Facebook bought Oculus Rift for $2 billion. "That just put everything into high gear," Kosslyn says. Facebook's move prompted Google to pump money into VR research. VCs started looking to make VR investments. Kosslyn and Thompson realized that VR was going to become another platform for business—and another way for startups to build on a powerful global capability, reach millions or billions of users, and take on companies entrenched in the physical world.

I liked their ideas, so I funded them to build a company, Angle. The company is creating tools that allow anyone to quickly and easily build apps or even a business in a virtual world. "One person with an idea should be able to do this," Kosslyn says.

VR has become good enough to make you feel like you're in another place, like a far-off city or aboard a spaceship. An Oculus adventure in 2017 still looks like a videogame, but there is an avalanche of money and talent falling into the space. By one estimate, startups and major companies such as Facebook, Microsoft, HTC, and Google spent more than $2 billion developing VR in 2015, and the advances are coming quickly. "What I thought would take 10 years got condensed into something like one or two," says Eugene Chung, who left Oculus to launch another VR startup, Penrose Studios.

Philip Rosedale, founder of VR company High Fidelity, is working toward creating a VR internet, linking VR worlds so we can move from one to another like we do today on the web. VR would then become less like a self-contained videogame and more of a global universe of content, community, commerce, and work. By the mid-2020s people will be able to choose how much of their lives to spend in the real world and how much in a parallel virtual world.

Augmented reality is halfway between virtual worlds and the real world. It merges the two together so digital information or images get layered into the physical world through, say, digital glasses or a smartphone screen. AR might even have a bigger impact than VR, and it's also a harder problem to solve. In early stages you might point your phone at a wall in your house and see what different paintings might look like on it, or you could point a phone at a corner in a city and see an overlay of what it looked like a hundred years ago. In 2016 Pokemon Go gave millions of players a taste of AR with its game of projecting Pokemon characters onto real-world settings.

Magic Leap, Google, Snap, and a few other companies are developing a more advanced AR that works through glasses that let you see both the real world and the AR projections. In early demos you might see a realistic-looking R2-D2 robot in your kitchen. Later the technology will let you sit at a conference table, put on a pair of AR glasses, and have a meeting with beautifully rendered full-size versions of your colleagues from around the world as they appear to sit in the other chairs.

I can think of some obvious ways VR and AR will contribute to unscaling. Sports could get unscaled. Instead of operating expensive major-league teams that need to build giant arenas to hold fifty thousand fans, niche leagues could cheaply assemble an audience that would experience the game in VR in a way that would never work in live sports. You might be right on the field for the whole game. In education, instead of going to a scaled-up college to sit in a classroom with other students and a professor, you could do it through AR or VR and get the same sense of community and collaboration.

* * *

Robots are becoming the physical manifestations of AI. The more they can learn and operate on their own, the more they will help drive unscaling.

Robots have been around for decades. We have robots on factory assembly lines, robots that pick stock in warehouses, robots that drill underground tunnels, robot Roomba vacuum cleaners. They're driven by software that has a fixed set of instructions. These bots can't really learn anything new—a Roomba might figure out a pattern to ensure it goes over every part of your rug, but that's about it. These kinds of robots don't unscale much of anything. Mostly they make large scale more efficient—like in a factory—and boost the economies of scale.

AI-powered robots will be a different story. For instance, it's almost mind-boggling what self-driving cars can do to the highly scaled global automotive complex. Traditional cars made us scale everything up. Every adult needed a car, even though that car would be parked 90 percent of the time. So more population meant more cars, and more cars meant scaling up car factories, building bigger and more highways, paving massive parking lots. The US interstate highway system began building out in 1956, and by 2016 there were 47,856 miles of federal highway.

As I write this, self-driving cars are advancing rapidly. Tesla cars can drive themselves already. Most major automakers are working on the technology. The cars will be connected to wireless networks. As more people buy networked auto-drive cars, they'll realize that it's silly to leave them parked most of the time. Why not let the car—by itself!—work for Uber or Lyft or some other on-demand car service? As more and more of these cars become available to the network, they will become a platform—and any entrepreneur in a garage could instantly create a nationwide transportation-as-a-service business by renting capacity on that platform. No need to buy a fleet of cars or trucks or spend a ton of money recruiting drivers—just configure the platform and go.

If robot cars become common, one car will be able to serve many people. Urban residents will be able to choose to rely on on-demand transportation because it will be cheaper and easier than owning a car, and niche companies will crop up to offer the kind of service needed. (Take your kids to school while you go to work? Take a whole football team to practices? Whatever you need!) Car companies will make far fewer cars in smaller factories. Highways will no longer need to be expanded. Parking lots can be turned into parks.

Drones are essentially flying robots. Next-generation drones will be infused with AI so they'll be able to learn. If drones are going to deliver Amazon packages, Domino's pizza, or rural mail for Canada's postal service—and each of these organizations is working on such plans—the drones will need to be able to navigate spaces, avoid people, and recognize and correctly react to things like dogs or electric lines. Emergency workers could send a fleet of AI drones into a flood zone to autonomously search for people who need help. AI drones might zip around a construction site, bringing workers parts and tools that they ask for by voice. I invested in a company, Airmap, that is building an AI-based platform that maps all the airspace on the planet and overlays the rules about every inch of it. Airmap technology is already implemented in systems at airports in major cities such as Denver and Los Angeles, and it powers an Apple Watch app that sends alerts about drone airspace. A drone can constantly talk to the Airmap database and know whether it's allowed to fly over this house or that university. It's an important part of making drones autonomous like cars—and safe as well.

Robot cars and drones will evolve into logistics platforms, once again taking a capability corporations used to build for themselves and making it easily accessible to entrepreneurs and tiny niche startups. Any startup will be able to log in to one of these platforms, configure it, and instantly have a way to move people or physical products anywhere on the planet. In a way, FedEx or a postal service is that kind of platform already, but robots and drones equipped with AI will be able to learn how to deliver specific items to specific places in the fastest and most efficient way—improving the ability to serve small markets profitably and better.

Other kinds of AI-driven robots will automate and unscale different kinds of work. You'll be able to hire a window-washing drone to buzz over your house and clean the outside of all your windows. Industrial robots already pull products off shelves in warehouses and will come to do the same at a consumer level—robot waiters or personal shoppers or clever small robots that you can hire to find something in your attic. I expect robot platforms to emerge across many different sectors, creating rentable automated labor that allows entrepreneurs to serve narrow markets, continuing the process of unscaling.

* * *

3D printing is currently transforming physical stuff into data, much the way digital technology over the past twenty years changed things like newspapers and phone calls into data. A 3D printer is a catch-all term for a robotic device that can take some kind of raw material—plastic powder, stainless steel—and shape it into a physical item based on digital blueprints. If your home computer printer takes a digital document and spits out a physical document, a 3D printer takes a digital design of a product and turns it into a physical product. Once physical items can be turned into data, that data can easily be sent anywhere over networks, making it cheap and easy for anyone to manipulate the design. And 3D printing machines can be connected together into on-demand, automated cloud factories that can efficiently make anything in customized small batches.

That's the idea behind a company I funded in 2017 called Voodoo Manufacturing, which was started by Max Friefeld, Oliver Ortlieb, Jon Schwartz, and Patrick Deem in 2015. The team began with a vision for the future of cloud-based manufacturing that will look a lot like present-day cloud computing. And Voodoo wants to become the AWS of cloud manufacturing. The company first built a factory full of 160 3D printing machines in the Bushwick section of Brooklyn, New York, all connected and supervised by smart software. Given the state of 3D printing technology, the machines could only make simple plastic products, but Voodoo found a market making parts for toys and promotional knick-knacks for marketing campaigns. "We built a system that lets anyone spin up low-volume manufacturing very quickly," Friefeld says. Unlike with mass production, an on-demand 3D printing center can manufacture one item for nearly the same cost as making one of one hundred thousand or 1 million. It can make those items as they're ordered—no need to predict demand and make and ship thousands of a product that may not sell. This, Friefeld states, will turn economies of scale upside down. "We're trying to reverse two hundred years of evolution in manufacturing," he says. Eventually, he believes, most companies that make physical goods will rent manufacturing as they need it, just as today they can rent cloud computing power as they need it.

Today and for the next few years 3D printers won't be able to make complex products. But that too will change. It's not out of the question that, for instance, 3D printers will be able to make a good sneaker. Consider what that might mean for the athletic shoe industry. Today Nike manufactures most of its shoes in China, Indonesia, and other Asian countries. This makes sense because labor is a huge part of the cost of making a shoe, and labor is far cheaper in much of Asia than it is in Western countries. To achieve economies of scale Nike operates huge factories that churn out shoes in anticipation of demand and ships them to retailers all over the planet, and the retailers then sell some shoes to customers and throw out the rest. In this model the enormous waste and transportation costs are worth it.

Now consider how that model changes if any shoe could be economically printed in, say, twenty minutes by some entity like Voodoo. Stores would become showrooms with no inventory. No shoe would be made until it's ordered. An unscaled shoe company could focus on design and marketing and then rent 3D printing operations to make the finished products. Because 3D designs could be altered as easily as we now change typefaces on a PowerPoint slide, customers could create their own style of shoe before it's made. Such is the promise of "distributed manufacturing." The World Economic Forum in 2015 named it one of the most important technology trends to watch. It is expected to have a mighty impact on jobs, geopolitics, and the climate. Take away the cost of labor, for instance, and that ends the most significant reason for outsourcing manufacturing to other countries. More products would be made inside the countries where they're sold, which would cut down on the energy burned to ship stuff all over the world.

However, it won't bring back manufacturing jobs. These on-demand manufacturing centers will be able to make products with little human intervention, operated instead by AI-driven software. In fact, the future of manufacturing suggests that today's big manufacturing countries—China, in particular—could face a crisis of job losses as concepts like Voodoo's catch on.

Meanwhile blockchain technology is contributing to unscaling. Blockchain is a sophisticated distributed ledger that keeps track of things on thousands or even millions of disparate computers, all constantly updating one another to ensure there is just one authentic digital version of anything recorded in the blockchain. That's why money, such as Bitcoin, was this technology's starting place. When you make a cat video, you want as many people as possible to copy it and pass it on. When you create money, it's good to make sure that when one person gives it to another, the giver can't keep a copy.

As blockchain develops, instead of having an internet that puts information and content online, we'll get a system that essentially automates trust and verification—the kind of stuff we now rely on accountants, lawyers, banks, and governments to do. You'll be able to know that anything on a blockchain (money, a deed, a person's identification information) is authentic. Better yet, because everything on the blockchain is digital, it is programmable. Currency can be programmed to keep track of every person who has used it. Software-enabled contracts can know if a job has been completed and make the payment without any middleman. A song on the blockchain could ask you to pay for it before it plays, cutting out iTunes or Spotify and sending the money back to the artist.

So blockchain is another form of automated commerce. The software does what offices full of people and traditional institutions used to do. Entrepreneurs can tap into the technology instead of building these capabilities themselves—another way to rent scale.

Everledger, for instance, is putting diamonds on the blockchain. First, Everledger's software creates a digital fingerprint of a cut diamond by measuring forty points on the stone. No two diamonds are exactly alike, and this creates a unique digital fingerprint. From that point on, the blockchain has an unalterable record of a diamond's path. If you can't trace a diamond back to a legitimate origin, you can assume it might be a diamond that funded a war or was stolen. Another blockchain company, Abra, changes how cash gets sent to individuals around the world. On one side are people who sign up, in an Uber-like way, to be virtual bank tellers. On the other side are users—like an immigrant in the United States who wants to send money to his mother in the Philippines. The user pulls up a map-like app to find the nearest teller, and the two agree to meet. The user gives the teller money, and the teller uses his or her account to put that amount of money into Abra's blockchain-based system. In the Philippines the user's mom similarly locates a teller, who translates the money into local cash to hand to Mom. The whole process cuts out banks, costs a fraction of the fees banks charge for such transfers, and can happen in an instant instead of ten business days.

In 2016 IBM started offering blockchain technology for supply chains. As more networked sensors get embedded everywhere, these devices will be able to communicate to blockchain-based ledgers to update or validate smart contracts. This would allow all parties to know whether the terms of a contract are met. For example, as a package moves along multiple distribution points, the package location and temperature information could be updated on a blockchain. If an item needs to be kept within a certain temperature range, everyone in the supply chain could know if that range was violated and exactly when and where it happened. That can also change the way businesses along a supply chain get paid. Smart contracts will automatically release payment as soon as goods have been delivered.

If you add up all these activities, the result is digital commerce platforms that allow even a one-person startup to compete against highly scaled companies. This is another way to put more of the "real world" into software, making it easier for an entrepreneur to set up and configure a global business or the kind of global supply chain that used to be the exclusive domain of big corporations.

* * *

In February 2001 the Human Genome Project and Craig Venter's Celera Genomics published the results of their human genome sequencing within a day of each other. The results were a 90 percent complete sequence of all 3· billion base pairs in the human genome. Venter later was quoted, saying his project took twenty thousand hours of processor time on a supercomputer. Getting that first sequence proved as daunting a project as putting the first man in space.

Now, less than two decades later, a company I invested in, Color Genomics, is offering a $249 genetic test that can sequence most of the pertinent genes in the human body. The goal is to make genetic sequencing so cheap and easy that every baby born will have it done and the data will inform his or her healthcare for life.

Across the medical landscape genetic data and AI are driving a shift to precision medicine—a shift, in other words, from mass-market medicine to market-of-one medicine. Healthcare will focus on each individual's body rather than standard practices. Companies like Color Genomics will give us all enormous amounts of data about our genetic makeup, while all sorts of devices collect other information, whether it's a Fitbit gathering data on our heart rate and exercise or a Livongo device monitoring glucose. AI and data will change medicine from prescriptive to predictive: doctors will be able to treat diseases such as cancer before they even manifest. We'll be able to figure out what makes you most healthy and build a new medical industry around keeping people well instead of treating the sick.

Imagine how this could unscale the pharmaceutical industry. Over the past fifty years the industry's goal was scale. Every company sought a "blockbuster" drug—a drug that would have some impact on the most people possible. Humira for arthritis, Crestor for cholesterol, and Viagra for erectile dysfunction were classic blockbusters. To find such a drug meant scaling up labs to test millions of substances, scaling up factories to make billions of pills, and scaling up marketing and advertising to convince millions of people they need the drug. And yet, because every human body is different, a blockbuster drug may or may not work for everyone, works differently in every body, and might be poisonous for some people.

Data changes that. It can reveal exactly what substances will work on you so a drug could be concocted just for you instead of for millions of people. Now imagine a drug industry built that way—where startups use data platforms and contract manufacturing to zero in on people with a particular disease and then custom-make drugs for each person. In an unscaled economy such small companies would be able to run a business like that profitably. It wouldn't need to build factories or spend years funding lab projects. Regulatory agencies, instead of approving each drug, would instead approve the process, making sure a company's data-driven approach will always make drugs that are safe.

AI-driven medicine will allow doctors to tailor care to every patient and focus on preventive and predictive medicine, which will drastically reduce the number of people who need to be in hospitals or even to see a doctor. Medicine will go back to feeling as personal as the days when the small-town doctor knew your family and made house calls. Entrepreneurial companies and doctors will be able to profitably compete against scaled-up hospital companies by focusing on individual patients or niche markets.

Genetic-based technology will have an impact beyond human healthcare. Startups are using synthetic biology to make new materials in small batches—imagine a small, local operation being able to make plastic out of microorganisms instead of a huge factory making it out of oil. Genetic engineers will make crops that thrive on city microfarms, unscaling food production and corporate farms and making it profitable for small, entrepreneurial operations to feed hyper-local markets.

All in all, genomics allows us to apply the rapid improvements of Moore's Law and the magic of AI to biology. Much as IoT is generating data and knowledge about nature and inanimate objects, genomics will give us data about life. If we have data about life, we can understand it, manipulate it, and program it on a micro-level. Economies of scale address mass markets. Economies of unscale prevail when entrepreneurs can address micro-markets. Genomics will drive the unscaling of healthcare, farming, and anything involving managing life.

* * *

This is the year 1900, the dawn of the twentieth century, times ten. We are reinventing our planet and ourselves. AI plus genomics will mean that precision health beats mass-market population health. AI plus 3D printing will help focused, niche production beat mass production. AI plus robotics will upend today's transportation system. AI plus VR and AR will recreate media and personal interactions. All these technologies will come together to revolutionize industry after industry, reversing a century of scale and driving unscale. This will also be a time of turmoil and opportunity. "Humanity is now entering a period of radical transformation in which technology has the potential to significantly raise the basic standards of living for every man, woman and child on the planet," write Peter Diamandis and Steven Kotler in *Abundance: The Future Is Better Than You Think.* They believe the new technologies will relentlessly drive down costs and make products—and our lives—better.

But great technological change can be difficult for people. "A society that had established countless routines and habits, norms and regulations to fit the conditions of the previous revolution does not find it easy to assimilate the new one," writes Carlota Perez. "So a process of institutional creative destruction will take place." As always, creative destruction is kind to the creators but brutal on those whose companies, careers, and finances get destroyed in the process.

All the new technologies raise difficult questions, and policymakers must pay attention and make sound choices. AI and robotics will wipe out millions of jobs—truck drivers, security guards, and delivery workers are but a few kinds of jobs that are about to get automated away by AI, robots, and drones. The automated accounting and banking in the electronic commerce platforms like Stripe will leave behind millions of finance professionals and contract lawyers. New manufacturing based on 3D printing promises a massive shift of jobs away from factories in China or Bangladesh and back to on-demand manufacturing shops in US and European cities. This can't be ignored. Policymakers need to grapple with how to help people make the transition to unscaling.

The mastering of human genomics will force us to deal with profound issues. Inventions like gene-editing technology CRISPR allow us to alter genes and thus alter people. We're close to being in control of our own evolution. I can foresee a startup eventually offering gene editing—customers will be able to buy themselves an upgrade, like

maybe thicker hair or better memory. If that comes to pass, we risk
creating a biological divide far more damaging than the old digital divide. Wealthy people will have the opportunity to make themselves
better, healthier, and smarter than poorer people, creating a gap between rich and poor that's not just about wealth and opportunity but
about talent and physical prowess. The impact on society would be
devastating.

From where I sit today it's possible to look ahead at unscaling, AI,
and other amazing new technologies and see opportunities in many of
the most important sectors of the global economy. In sector after sector
demand, once aggregated by big companies, is getting unbundled and
addressed by small companies, which sometimes then get big by reassembling a new cross-section of customers in a new way. This cycle of
taking apart and rebundling a market in an innovative way is happening faster and faster as scale becomes cheaper and easier to rent and as
software and data create insights that lead to the reinvention of product after product.

This is where the idea of unscaling meets reality—where you as an
entrepreneur or investor or individual can discover your path in this
new economy. That's what you'll find in the next section: a look at
how major sectors will get reinvented and what that means for all of us.

PART 2

THE GLOBAL REWRITE

3

Energy

Your Home Will Have Its Own Clean Power Plant

Solving climate change is one of the great opportunities for entrepreneurs of our time. And artificial intelligence and the economies of unscale are giving innovators new ways to create alternatives to burning fossil fuels. In this sense unscaling might actually help save the planet. But there's a catch: heavy regulation in energy—particularly the electric power industry—means change can be devastatingly slow to come.

Naimish Patel, founder of energy startup Gridco Systems, knows all of this very well. His epiphany about the opportunity in energy tells us a lot about how the energy industry can evolve, whereas his struggles to deal with a regulated mentality show the challenges ahead. In 1998 Patel helped found Sycamore Networks, which made optical switches and software that helped move data around fiber-optic communications networks, and he worked as its chief technology officer. The company was red-hot during the late-1990s public infatuation with the first generation of internet companies, rocketing to a $45 billion valuation in 2000. The bubble then burst by the end of that year, and the sense at the time that the internet had been overhyped dragged down

49

the value of telecom companies, including Sycamore. Patel left the company in the mid-2000s, and while he looked for his next opportunity I brought him in as an entrepreneur-in-residence at General Catalyst, giving him the time and funding to find a new venture to start.

Patel made a trip to Iceland in 2007 to help set up a data center. While Patel was there to advise about communications he came to realize that the biggest cost for large data centers was power. This center was going to be built in Iceland to take advantage of the cloudy and chilly climate to keep it cool. The air pumped in from outside the building was like free air conditioning. "I got to know the power service providers while I was there," Patel now says. And he discovered that "many of the things I took for granted in communication weren't there in power." The systems were technologically far behind, with little software or digital electronics automating the flow of power, monitoring it for efficiency, or instantly rerouting power when something went wrong. In short, the power grid worked like a river—a constant one-way flow that was barely managed. Patel thought it could work more like the internet, allowing software and switches to instantly route power from anyone to anyone. "I started asking if there was an opportunity to bring that automation to the electric power system," he says.

As Patel studied the industry, he saw that customers were starting to demand an unscaled approach to power that utilities couldn't meet. Customers were putting solar panels on rooftops and sending power back into the grid—something grids, in their river-like state of operations, weren't designed to handle. Customers were installing connected gadgets to better control power usage. That, along with new products and services such as Tesla's introduction of its first electric car model, threatened to alter the expected patterns of power usage. If millions of people charge their cars all night, they'll spike energy consumption during a time when utilities expect consumption to be low. Power grids weren't built to easily detect or respond to that kind of shift in customer behavior.

"I saw the beginning of customer-driven change," Patel says. "One-service-fits-all is no longer going to be sufficient." With that observation he founded Gridco, a company that would make internet-style switches and software for the electric grid. The internet can transmit information in two directions—from any provider located anywhere to users and then back—because of a system of routers and the software inside those routers. The twenty-first-century power grid is going to need to work in a similar way, allowing power to be conveyed where and when it is needed in the most cost-effective way and allowing any user who might be generating power to sell it back into the system. Patel calls it an "active grid infrastructure."

Gridco is a small company that may or may not be the one to overhaul the power grid, but some company like Gridco will, without a doubt, usher deep innovation into the energy sector. One way to explain why is to look at the history of computing. In an older era, when all computing was on big mainframes in corporate back offices, only specialized professionals could write software and implement new ways to use computers. As computing got democratized and distributed—first with the arrival of personal computers and later with cloud computing over the internet—almost anyone could create new applications and products. The result has been an exponential boom in the uses of computing. Now imagine that the legacy electric grid is like a mainframe computer—a closed, inaccessible system that only certain professionals can connect to and modify. In the thinking of innovators like Patel, the generation and transmission of power can be taken down a path similar to PCs and then cloud computing, gradually opening up and democratizing energy generation and transmission—allowing entrepreneurs to think up new ways to generate, manage, and market power while plugging into the grid as easily as a PC taps into cloud computing through the internet. Just imagine how the democratization of computing has opened possibilities for innovators to create services and applications. An internet-like, AI-governed electric grid would present similar opportunities, opening up vast opportunities for startups and new products and services that might include small-scale private solar farms (or, someday, even small nuclear reactors, which I'll get into later in this chapter) or differentiated services that sell higher-reliability power to businesses at higher prices and power that comes with fewer guarantees to individuals at deep discounts. No one is sure what might transpire because an open, internet-like grid is still such a new concept.

As of this writing Gridco is straining against an industry set up to resist innovation. Gridco closed a $12 million round of financing in 2016. That's a small amount when compared with, say, the billion-dollar rounds closed by companies like Uber. In fact, because of regulation and the sector's resistance to innovation, energy doesn't attract as much capital as it should. So with Gridco it's hard to know whether the company itself will live on, but the company is symptomatic of the entrepreneurial thinking that will change the industry's mindset.

Much of the conversation around fighting climate change gets hung up on austerity. Opponents especially paint it as some mix of sacrifice and grudging duty at best, ruinous to the economy at worst. US president Donald Trump brought such a viewpoint to his decision to back out of the Paris Accords, which aimed to set limits on carbon emissions.

But energy entrepreneurs take a more positive approach. They see the opportunity for prosperity and for helping innovators take advantage of global momentum to reduce carbon in the atmosphere. But it will require some changes in how we regulate and govern energy as well as some changes in attitude among financiers as they come to realize energy startups can bring a big return on investment.

Just as we've seen when internet companies redefine sectors like communications, retail, and media, newcomers in energy will bring fresh approaches that can improve the efficiency of the power grid and replace carbon-based energy with clean, advanced-energy technology. If the internet's past tells us about what will happen as energy democratizes, new companies will create jobs and boost the economy. Curtailing carbon will be a net plus for the global economy.

* * *

In 1900 roughly half of the energy consumed by the world came from burning biological material such as wood, corn stalks, and dried manure. The other half came from burning coal, usually for heat and, increasingly, to turn electricity-generating turbines. Trains and ships ran on coal. This energy equation drastically shifted with the rapid and widespread development of cars, trucks, and airplanes over the course of a century. By 2000 almost all the world's energy was generated by burning oil, coal, or natural gas; hydroelectric and nuclear power and burning biofuels played a much smaller role. Along the way, global per capita consumption of energy more than doubled, dominated by soaring consumption in developed countries. There were much smaller increases in the developing world, but as economies improve and raise the prosperity of more people, those people also tend to use more energy.

As demand mushroomed, we addressed it with scale. The world quickly concluded that the economics of energy did not work on a small-business scale. It cost too much to drill for oil or gas, ship it, refine it, and get it to market in small batches. The same was true for generating electricity, maintaining a grid, and servicing customers. The bigger the energy company, the more it could benefit from the economies of scale.

We built enormous power plants so one plant could generate electricity for thousands of homes. We built one-size-fits-all electrical grids that delivered the same power in the same way to every home and business. We scaled up an oil industry that embraced scale and mass markets. Some version of ExxonMobil has been one of the biggest companies in the world for all of the *Fortune* 500's existence. (The first *Fortune* 500 list, in 1955, placed Jersey Standard—ExxonMobil's name then—at number two, behind General Motors and just ahead of US Steel.)

Every facet of twentieth-century energy, all over the world, was about getting to the biggest scale possible. To get to the biggest scale possible in electric energy we intentionally created local monopolies to drive economies of scale. What sense would it make to have two competing grids next to each other? The monopolies could soak up *all* the demand from a region. The trade-off for allowing an electric utility to become a monopoly was that it had to agree to be regulated.

A sector that is so thoroughly scaled, monopolistic, and ruled by regulations and governments has less incentive to innovate, invest in renewable energy, or more efficiently and effectively serve niche markets instead of mass markets. Monopoly utilities are wasteful because they can be—they get paid on a fixed-rate-of-return model in which revenue is guaranteed to be some standard percentage above their capital base. There is almost no *economic* incentive for a scaled energy industry to solve the climate change problem it is creating.

That's not necessarily the fault of energy executives; it's how the system was built. Regulators focus on egalitarian access to energy and on reliability—in so many ways a good thing. The approach in the developed world keeps the lights on nearly 100 percent of the time. But that approach discourages risk and incentivizes safety and predictability. So developed nations wind up with a highly reliable, risk-averse, and wasteful energy industry—not a great model for aggressively addressing climate change or responding to fast-changing demands for power, as with the emergence of electric cars. In developing nations such as India, where economies and populations are booming, Western utility models often break down. It can cost $1 billion and take years to build a new power plant, and poorly maintained grids often prove unreliable. A more entrepreneurial, distributed model of power production and transmission could encourage entrepreneurs to set up solar or wind production wherever it's needed, sell excess back into the grid, or pull power from the grid when demand spikes. All around the world the market for power is transforming, and the energy sector will need to be nimble and innovative enough to transform with it.

* * *

Energy and transportation go hand in hand. Energy made the transportation complex possible, and transportation created massive demand for energy. Just as the energy sector sought to operate at scale, so did transportation. Governments built highway systems and airports. Car manufacturers consolidated into a handful of global giants. Airlines and shipping companies gained advantages by getting as big as possible. Scaling up this energy-transportation system had enormous benefits for society, making people much more mobile and changing the course of history.

Now, though, the highly scaled energy-transportation system threatens our planet. Scale in energy and transportation was the right answer for the past one hundred years. But today, when we must combat climate change, we need to reverse course. We need to unscale energy and transportation so entrepreneurs can innovate and serve markets more efficiently—just as is happening in other industries such as retail and media.

The scaled approach built *inefficiency* into the system. The electric grid and highways were overbuilt to serve the most people at peak times, even if those resources went to waste most of the rest of the time. The car is perhaps the most tangible example of this wanton inefficiency. We waste a vast amount of energy building and maintaining a car, often two or more cars, for almost every person—yet most cars stay parked and idle about 90 percent of the time. If one result of unscaling is creating focused, innovative companies around car sharing and on-demand transportation, we'll expend less energy to build fewer cars that will serve more people, greatly improving energy efficiency while remaking transportation into a customizable service that's actually more practical than owning a car. After all, most people just want to get from point A to point B. Owning a big, expensive machine has long been the best way to do that in most places, but that won't necessarily be the case in coming years.

Though energy and transportation are, separately, giant sectors that will go through profound changes as they unscale, it's impossible to pull them apart because of their impact on each other. Energy can't unscale without transportation unscaling, and vice versa. Transportation will essentially become a rentable service built on top of modern energy platforms—that would be one way to think of a self-driving Uber-like car service that can be summoned on a smartphone. The thinking of entrepreneurs, incumbent-company CEOs, regulators, and lawmakers needs to and will change across both industries. Tesla CEO Elon Musk is one entrepreneur showing the way. Tesla, founded in 2003 in California, started out by bringing a high-performance electric sports car to market and has since expanded into electric sedans, home battery systems, and solar power. Musk gets it right when he thinks about his pioneering company as an integrated energy and transportation entity. Musk in 2016 put forth his "Master Plan, Part Deux" (a sequel to his first Master Plan, which he published in 2006). He wrote that Tesla's ultimate goal was never to produce hot electric cars (even though Tesla unveiled the fastest-accelerating car on earth). Tesla built hot electric cars as an entry point for ending dependence on oil. "The point of all this was, and remains, accelerating the advent of sustainable energy, so that we can imagine far into the future and life is still good," Musk wrote. Tesla cars will be a part of a sustainable electric system that includes solar panels, batteries, and software to manage power and trade it over networks.

In 2017, just as Tesla started selling its $35,000 Model 3, the transition to electric cars accelerated faster than Musk or almost anyone would've thought. Volvo Cars became the first mainstream automaker to announce that all the models it introduces starting in 2019 will be either hybrids or powered solely by batteries. France, a major auto-making nation, set a goal to end the sale of gasoline and diesel cars in the country by 2040. India got even more aggressive, setting a goal of selling only electric cars in the country by 2030. At the same time, entrepreneurs are experimenting with ways to get more cars off the road. A Finnish company, backed by Toyota, is promoting what it calls "mobility as a service," or MaaS. The company, MaaS Global, developed a transportation subscription service called Whim. In 2017 early users paid from around $100 to $400, depending on the level of service, for Whim. A user could select a destination on a map, and the app would list possible ways to get there, including taxis, public transportation, rental cars, and bikes—choose the best one, and the subscription fee covers the cost. Founder Samp Hietanen started the company based on a research paper he wrote while working for a Finnish smart-transportation think tank, ITS Finland, and Hietanen wants to take the concept global. "What if you had 'unlimited Europe'—ground and air transport included—through one app that would be your companion wherever you go? Then you could truly be a global citizen," Hietanen told reporters.

Policymakers and regulators at the national and local levels need to embrace unscaling and the efforts of entrepreneurs like Hietanen, Patel, and Musk. If we can get out of the innovators' way and set them loose, energy and transportation will look much different in twenty years.

* * *

A ten-year unscaled vision for energy and transportation goes like this:

An increasing number of homes and buildings will have cheap and super-efficient solar panels on their roofs and high-powered batteries in their basements or garages. The batteries will store power generated when the sun shines for use when it doesn't. The electric grid will operate more like the internet, allowing anyone to sell excess energy or buy needed energy in an eBay-style marketplace. Energy customers will have more choice about where their power comes from, much the way they can now choose many different ways to make a phone call (landline, cell phone, Skype, and so on).

A growing number of cars by then will be electric, and fewer people will need to own cars because on-demand transportation—perhaps Uber or Lyft self-driving cars—will likely be available in most densely populated areas. Home solar panels and batteries will charge a family's electric car so the home will supply much of the clean power a family needs. Some families might not need to buy any electricity from far-off coal-burning electric plants or ever again fill up at a gas station.

If these developments come to pass, utilities won't need to build large-scale, carbon-based power plants. They won't be necessary if more and more homes and businesses are generating electricity on their rooftops and the grid works more like the internet, moving power to where it's needed and storing excess in new-age batteries for reliable access. If Patel's Gridco and other innovators make the grid more like the internet, then just as anyone can now be a hotelier because of Airbnb, in a decade anyone will be able to be a mini power company once these new technologies are in place. In an unscaled era the power sources in a home or small business will likely be better, cheaper, cleaner, and more resilient than the next massive power plant. And if new energy technologies are better, cheaper, and cleaner than the old ones, that's what customers will choose—which should carve a path away from so much carbon burning worldwide.

The forward-thinking electric utilities will evolve to become platforms, essentially operating an energy version of the internet, much the way companies such as AT&T maintain telecommunications systems that others can build on. Entrepreneurial companies will build products and services on top of those energy platforms, selling services to customers with specific needs. (Maybe you'd want a package for powering your home and your car that allows you to recharge your car at stations away from home when traveling.)

As transportation moves to the electric grid and off carbon, oil will become a shrinking piece of the energy pie—much like the decline we see today in coal. Some people will drive gas-powered cars for fun, just as some people still ride horses. One by one, gas stations will go out of business or convert to electric car charging stations. As demand for oil drops and prices plunge further, drilling new wells will become bad business. Before long the amount of carbon we're chugging into the air should drop dramatically.

* * *

Since the early 2000s electric power companies have been talking about a "smart grid"—adding sensors and computing to the grid to monitor power flow and analyze usage. Congress and the US Department of Energy have been nudging utilities since 2007 to reinvent their grids to move them from closed systems controlled by the utilities to open systems that others can connect to. Startups like Gridco have been developing the switches and software that can help modernize the grid. Yet progress has been slow, in large part because of the risk-averse nature of utilities.

That's beginning to change, especially as consumers and businesses increasingly install solar panels that pump power back into the grid. Customers are demanding choice and control when buying and using energy. "The risk of continuing business as usual is immense in terms of system reliability and costs associated with inefficiencies—which many stakeholders in the electric power sector recognize and want to avoid," says Ignacio Pérez-Arriaga, visiting professor at MIT and an author of an MIT report, "Utility of the Future." The report, released in December 2016, is the result of a multiyear research study, and its conclusions are in line with those presented here. Other recommendations include dynamic pricing, so utilities can charge different rates in different places or at different times, depending on supply and demand, and designing the grid to make it easier for solar cells and batteries to plug into the system, no matter their size or voltage.

The grid, as noted above, is evolving into a platform. And as a platform it can become a key to unscaling: a rentable resource, like cloud computing, on which small, product-focused companies can innovate and address smaller niche markets. Call it the "power cloud"—an energy platform that will enable an explosion of unscaled energy businesses. Utilities that make this transition to a platform will actually become even more essential than they are today.

Solar will be a major driver of new unscaled energy business. The technology of rooftop solar is on a predictable trajectory—echoing Moore's Law, which explained why computers got twice as powerful for the same price every eighteen months for decades, though solar is not improving at that same breakneck pace. Still, the cost of solar has dropped 95 percent since the 1980s while efficiency has rocketed. And solar farms, with acres of solar panels laid out in a sunny climate, can be the most efficient of all. With today's technology solar panels covering just a few counties in Texas could supply all the electricity needed for the entire United States. By some calculations the amount of solar energy that hits the earth is more than five thousand times the amount of energy all of humanity uses. The challenge has been harvesting it. Solve that problem, and mankind would never need to burn another atom of carbon.

In 2000 Germany put into force its Renewable Energy Resources Act, which mandated a rollout of solar energy in certain parts of the country. By 2017 solar was accounting for 7.5 percent of Germany's electricity needs, allowing the country to start decommissioning nuclear power plants. China's government has aggressively invested in renewable energy technology. It pumped $89.5 billion into the industry in 2015 alone, and in January 2017 its National Energy Administration announced it would spend $360 billion through 2020 on renewable energy. With its cities choking on smog and pollution, China desperately needs to move away from carbon-burning power. Meanwhile Walmart, which has been installing solar panels on the rooftops of its stores, announced it intends to operate on 100 percent solar. (It's not all that close yet, but it committed to cutting energy use off the grid by 20 percent from 2010 to 2020.) And Google, which in 2016 consumed as much energy as the whole city of San Francisco, said that by the end of 2017 all its data centers globally will run entirely on renewable energy. Efforts like those of Germany, China, Walmart, and Google drive up demand and create a market for innovative and inexpensive solar. At the end of 2016, for the first time, solar was a cheaper source of electricity in many parts of the world than any other energy technology, according to data gathered by Bloomberg New Energy Finance.

I saw this play out firsthand. In the mid-2000s I invested in an American solar panel company by the name of Stion. People like me from the tech industry have an intuitive understanding of the solar business because the processes for making solar panels are similar to building computer chips. Technologists can constantly reduce the thickness of solar cells and use less silicon per watt, driving down manufacturing costs while increasing each cell's efficiency. Still, the manufacturing of panels was not getting cheap enough fast enough because demand was not growing enough to justify pumping huge amounts of money into solar R&D and advanced manufacturing. Solar was a chicken-and-egg problem. The Chinese muscled through that equation by mandating massive demand and then ramping up efficient factories to produce solar panels and drive down the cost. Our company could not survive the onslaught of inexpensive Chinese-made panels, and that proved costly to a lot of US-based solar companies. Stion planned to create a thousand jobs in Hattiesburg, Mississippi, but instead only created about 110. But China produced a net benefit for the whole world by making solar affordable.

The situation today bears similarities to computing in the 1980s, as personal computers finally brought the cost of computing down enough so anyone could buy one and use it to start a computer business. Solar has reached its PC moment. It's now economically viable for individuals to install panels and build their own power plant—and I expect it will soon get easier for any entrepreneur to start a solar power–generating business and connect it to the grid.

Wind power has a different dynamic. In renewable energy wind power is actually not unscalable. Wind has economics similar to that of power plants. Wind is designed for scale. The enormous windmills you see on hillsides generate enough power to make them worthwhile. But you can't put a small windmill on your roof and get enough power to make a difference; the physics don't work that way. If energy is trending toward unscale, wind is likely to play a relatively minor role in the future of energy.

Although solar will have the biggest impact on energy production, the new power technologies in transportation will have an enormous impact on demand.

We all know how far electric cars have come. An electric car was a pipedream at the turn of the twenty-first century, but by 2017 seeing a Tesla in the San Francisco Bay Area was no more remarkable than seeing a Toyota Camry. When Tesla took pre-orders for its Model 3 sedan in March 2016, it signed up half a million buyers within weeks. General Motors has embraced electric cars, declaring that they are its future. Most of the world's automakers have made similar declarations.

At the same time, Uber and Lyft have accustomed millions of people around the world to the concept of car sharing—or, if you will, on-demand transportation. Over the past fifty years the car companies convinced everyone that we had to own at least one car, if not more. Car sharing is breaking that cycle. It's teaching us how to get around without owning a car. One car can serve many people now instead of one person owning many cars.

The next step in car sharing will be driverless cars, which can be routed to users much more efficiently than human-guided cars. (After all, humans are human—some drivers will decide to turn down a request for a ride because maybe they want to go to lunch.) Now, it's harder to say *when* this will arrive than it is to say that it *will* arrive. As I'm writing this, Uber is offering rides in autonomous cars in Pittsburgh, and major car companies—Ford, Volvo, BMW—are predicting that they will be selling driverless cars by 2021. General Motors and Lyft are collaborating on self-driving cars and say they'll be ready by 2021, though Lyft CEO John Zimmer says the vehicles will serve limited geographic areas where top speed is twenty-five miles an hour. Tesla cars are already completely capable of driving themselves in many situations, though a driver still needs to be behind the wheel, ready to take control. But it's going to take years more before fleets of cars with no driver roam cities. "These statements are aspirations, they're not really reality," says Raj Rajkumar, a professor of engineering at Carnegie Mellon University, who collaborates with General Motors. "There's still a long way to go before we can take the driver away from the driver's seat." Mary Cummings, a professor of mechanical, electrical, and computer engineering at Duke University, says a fully autonomous car that "operates by itself under all conditions, period [is] a good 15 to 20 years out."

Yet if we play out the trend, at some point fleets of auto-driven cars will be roaming every city, and then owning a car will seem like digging a well to get water—you might still need it in a rural setting, but not in urban areas.

If we replace many of our gas-burning cars with auto-driven electric cars, we will move massive energy demand away from carbon and onto the electric power grid, which will increasingly be fed by solar. The rise of that demand will create fantastic opportunities for new kinds of businesses in energy, drawing yet more entrepreneurs who will dream up yet more unscaled energy companies.

All this unscaling means the world's auto manufacturers are destined to eventually shrink and make fewer cars. That in itself will help the environment. Each car takes an enormous amount of energy to make. Imagine the manufacture of all the car's parts all over the world, shipping those parts to a Ford or GM plant, and the energy involved in keeping that plant running—and then shipping the cars to dealerships all over the country. Simply making fewer cars because each car will serve more people will greatly reduce the burning of carbon.

One of the keys to electric cars and to all of energy unscaling is batteries. Batteries will be necessary to overcome the greatest problem with solar-generated electric energy: there's no natural way to store sunlight. Oil can be put in a tank for when you need it. So can natural gas. Coal can be piled high. But sunlight—when it's gone, it's gone. No one wants to rely on solar or any kind of noncarbon power that can't be stored and might not be available when you need it. Viable home batteries have been a lagging link to making all this unscaling accelerate.

Elon Musk, as of mid-2017, was about to open Tesla's $5 billion Gigafactory in Nevada, and for the first time large batteries for cars and homes will be manufactured in bulk. Tesla has said its factories should drive down the cost of battery power by at least 70 percent. Meanwhile other companies around the world are working on new kinds of batteries. In Pittsburgh Aquion Energy has been developing what it calls saltwater batteries and has installed them on a solar farm in Puerto Rico. In the United Kingdom Dyson, the engineering company that makes the popular vacuum cleaner, is working on home batteries. In Germany Mercedes makes a home battery that it intends to roll out to global markets. John Goodenough announced in early 2017 that he and his team at the University of Texas at Austin had invented a glass-based battery that blows away the performance of every previous kind of battery, including lithium-ion batteries—which were invented in the 1980s by . . . him. Goodenough's new battery can store three times more energy than a comparable lithium-ion battery, according to the Institute of Electrical and Electronic Engineers.

Home batteries in the mid-2010s still cost thousands of dollars. They're difficult to install and don't hold enough power to make it through a handful of rainy days. But with so many companies seeing the opportunity, that will change. Consider what will happen once we get low-cost, high-performance home and car batteries. That's when each home or business can become its own power source, breaking free of the grid and of any dependence on carbon-based fuels. Highly efficient solar panels will generate electricity when the sun shines, storing all the excess in batteries in the basement or in the battery of the electric car parked in the garage.

That would be the ultimate unscaling of energy—instead of monolithic power companies serving all customers the same way, each individual building will make and use its own power the way its inhabitants want to. We'd go from a fragile grid that can be disrupted by a storm to a distributed grid that would be much more resilient. All in all, once the problem of storage gets solved, energy unscaling will accelerate into high gear, and old models of electric grids and oil companies will crumble.

One last but important piece of the energy-unscaling puzzle is data. Smart things started making their way into homes and commercial settings in the 2010s. On the consumer side Nest popularized the smart thermostat that can learn the patterns of a home's residents and use that information to more efficiently control heat and air conditioning. Smart lights from Philips and GE could gather data about lighting use and automatically turn on and off. In businesses General Electric introduced the "industrial internet," and tech giants such as Cisco and IBM created IoT offerings that could put sensors into almost anything that ran on electric power.

All this IoT activity is generating enormous amounts of data about energy usage—information no one ever had before. These insights will play an important role in unscaling. Just as data about retail transactions or social networks has helped entrepreneurs customize products and target small units of demand, the same will happen in energy. The data will guide innovators to new markets and new kinds of products in energy. It will help companies make better solar technology, better batteries, and better electric cars. Data will help policymakers better understand the energy landscape and shape smart regulations that help unscaling go forward.

As is often said, in the twenty-first century data is the new oil— the raw material that makes everything go. In energy, data is more than just a metaphorical new oil—it's critical to how we'll actually replace oil.

* * *

Around 2007 I started investing in energy. I was worried, though, about the industry's record of resisting innovation, so I learned to work on underlying policy to try to knock down some of the roadblocks to creating open power systems. I was then living in the Boston area, and Deval Patrick had just been elected governor. He encouraged me and other investors and entrepreneurs to develop ideas about how to design policy around unscaling energy and to help entrepreneurs figure out how to build businesses in the energy sector. I helped found the New England Clean Energy Council, made up of people from the energy industry, investors, academics, and policymakers. We helped develop legislation, including the state's Green Jobs Act of 2008, which provided funding and mechanisms to create clean-energy jobs.

Soon after, once I moved to Silicon Valley, I helped start a public policy organization called Advanced Energy Economy (AEE) in 2011. A cofounder was Tom Steyer, a hedge fund manager and political activist, whom I met through mutual friends. George Shultz, the former secretary of state, joined the board, as did former Colorado governor Bill Ritter. Now AEE is in nearly thirty states, focusing on the next-generation electric system and how to align regulations with business models to support innovation. Because of my work with AEE, I became intimately informed with the interaction of policy, technology, and finance and how we need to think about all three together as a system. That will be particularly necessary in energy and transportation. As unscaling takes hold, policy will need to change in smart ways; otherwise, antiquated policies will either stifle unscaling or allow unscaling to happen without enough guidance.

Look at just one example of how regulation unnecessarily gets in the way of unscaling. In 2016 regulators in Nevada gave in to pushback from the incumbent power company, which was feeling a financial crunch. As a result, the price paid to homeowners who sell excess energy generated by their rooftop panels back to NV Energy dropped significantly. NV Energy had been reimbursing solar customers about 11 cents per kilowatt hour for the excess energy they generated, and that was cut to about 9 cents and will drop to 2.6 cents in 2020. Solar providers said the new rates make solar too expensive, discouraging homeowners from installing panels. The solar industry and the state's Bureau of Consumer Protection have been challenging the rate change. Governor Brian Sandoval said he wants to support solar but that the state needs to find the right balance of subsidy for the solar industry. The price drop, though, was so damaging to solar customers that the nation's biggest rooftop solar company, SolarCity (part of Elon Musk's empire), decided to pull out of the state, laying off 550 workers and setting back solar energy in a desert state that should get much of its energy from solar. Entrepreneurial companies like SolarCity tend to unscale energy, while monoliths like NV Energy work to maintain scale.

As policymakers think about unscaling, they'd do well to learn from the way many nations rethought telecommunications regulations at the dawn of the internet. The regulatory changes helped accelerate innovation in communications and create more open networks.

The US Telecommunications Act of 1996 was the first major regulatory change for that sector since the Communications Act of 1934. The 1996 law created a clear distinction between telecommunication providers and information service providers, legally absolving the latter from common carrier requirements. So broadband service providers—for instance, cable companies and wireless broadband providers—were set free from being required to make sure everyone had a phone line. That requirement was much like the mandate that makes electric utility companies make sure everyone's lights stay on, which of course is a good thing for society. And yet companies that must give us reliable and affordable service for all are necessarily mass-market, highly scaled, one-size-fits-all providers. Companies released from requirements such as this can, in addition to giving us reliable service, focus on tailoring services to smaller niches—which is the heart of unscaling. A common carrier tends to offer all customers as little choice as possible. Allow for more innovation, and hundreds or thousands of entrepreneurial companies can build services for small groups of customers, in essence creating an array of choices for each customer. And, again, putting customer choice at the center drives unscaling.

One way or another, energy demand is going to shift and customers will produce more of their own power. In fact, if utilities can't respond to changing demand and pricing, customers will have the incentive to further adopt unscaled generation or energy (which is likely to get cheaper and have more interesting features) instead of utility-generated energy (which is likely to get more expensive while offering one-size-fits-all services). As customers peel away from utilities, fewer customers will be paying to keep the utility's system going, so, in turn, utilities will need to raise prices further, ensuring that even more customers will leave. Its assets—power plants and grids—would be left stranded and losing money. Some have termed this phenomenon the "utility death spiral."

The swift change to electric-powered transportation makes it all the more important to transform the twenty-first-century electricity system. A crucial early step would be to align utilities *with* entrepreneurs instead of *against* them. Governments need to create regulatory and business models that give utilities a path to evolve their role from operating mammoth power plants and transmission grids to operating software-powered platforms that interconnect the small-scale power solutions in homes and small businesses. The grid can then become a platform that supports entrepreneurial solutions, much like the internet or the iPhone and app store. Utilities would be able to prosper, providing reliability and resiliency in the power network while faster-moving entrepreneurs create ever more effective ways to generate, move, trade, market, share, and store power.

Time after time we're seeing small companies win in established sectors by building on existing platforms and finding a new market. That's what needs to happen in energy. If the right technologies get developed, entrepreneurs will unscale the energy industry, turning it into an ecosystem of small companies and producers generating and moving power in a distributed way.

＊ ＊ ＊

The United States can create jobs in energy by allowing world-class innovative energy startups to flourish. The United States is not likely to beat China at low-cost solar manufacturing, however, just as it did not beat China at low-cost computer manufacturing. But in energy, much like in the internet industry, the United States can unleash entrepreneurs to create products and services built on top of commodity energy platforms.

As an investor looking at the horizon, these are some of the opportunities I see coming for entrepreneurs and established companies:

REBUILDING THE GRID: The power grid is an antiquated one-way system reliant on a lot of aging infrastructure. The grid needs to be rebuilt into a two-way internet-like system that allows energy to flow from any producer to any consumer. It needs to be an open system like the internet so anyone can build applications on top of it—whether those apps are electric cars, smart homes, or new kinds of businesses no one has thought of yet.

At the beginning of this century US cable TV companies rushed to rebuild their one-way broadcast networks into two-way broadband internet systems. It was a huge undertaking that generated business for boots-on-the-ground cable rebuild companies, internet hardware companies such as Cisco and Naimish Patel's Sycamore Networks, and all kinds of networking software companies. Rebuilding the electric grid will be a task that is ten times bigger than building broadband. The physical rebuild—replacing transformers, power lines, and equipment all over the country and all over the world—will also create jobs in every town it touches. How much money will be on the table? The American Society of Civil Engineers is calling for an additional $177 billion in spending on the grid by 2025.

INTERNET OF THINGS AND THE "POWER CLOUD": As the grid gets rebuilt into an open two-way system, it will open up the market for products and services that can get built on top of that platform—much as cloud computing opened up possibilities for connected devices like a smartphone or Fitbit and services ranging from Uber to Salesforce.com. We'll see devices that measure and regulate power in homes and buildings, giving much more detailed views of how power gets used. (Do you really have any idea how all the electricity you pay for is getting used? Probably not!) Cloud services will pop up that will allow customers to buy power from any provider, or they can aggregate power and sell it to others.

It's hard to predict the clever ideas we'll see from entrepreneurs, but we can get a glimpse by looking at some of the startups in this space. LoudCell, for instance, can deploy sensors and software to give companies a dashboard so they can see their energy consumption, generation (if they have solar panels feeding energy back into the grid), and wastage. A company called Bastille is developing IoT devices and software that can monitor an electric grid and detect threats from hackers, storms, or anything that might disrupt service.

ELECTRIFIED TRANSPORTATION: Cars and trucks, without a doubt, will increasingly be electric. Over time ground transportation will move off of carbon and onto the grid. Imagine the enormous opportunity to build charging stations or to solve the biggest impediment to electric transportation: long-distance travel. Every gas station will need to be transformed. Parking garages at companies and in cities will likely be fitted with charging stations.

ChargePoint is already rolling out its ExpressPlus charging stations, which are supposed to let electric vehicle drivers add hundreds of miles of driving range in the time it takes to stop for coffee. "By the time your latte hits the end of the counter, your car's probably charged. That's what we want to get this to," CEO Pasquale Romano told *Forbes*. "It can't be any harder than going to a gas station." Tesla has a competing product that it says can replenish 170 miles of range in its cars in about thirty minutes. Expect an explosion of activity in this market as electric cars populate the roads.

THE STORAGE CHALLENGE: Battery technology remains one of the toughest problems to solve in moving power off carbon and over to solar or wind generation. Oil and natural gas can be stored in a tank for use later. The only way to store sunshine or wind is to hold the power it generates in batteries. But batteries aren't yet good enough or cheap enough to universally solve this problem. We'll need batteries that can store enough solar-generated power to last through a week of stormy days or power a car for a full day of driving. Getting there will take breakthroughs in materials science. The company that gets there first will change the world.

Jay Whitacre, of Carnegie Mellon and Aquion Energy, is pushing ahead with his "salt water" battery. A Chinese company, Contemporary Amperex Technology Ltd (CATL), is competing against Tesla to become a giant manufacturer of lithium-ion batteries, the type of batteries that now power electric cars and serve as home batteries. Because of the hard science and big manufacturing involved, this will be a tougher sector for startups. Yet venture investment is flowing to companies.

NEW NUKES: Somewhere down the road lies the promise of energy-generation technology that will solve all our energy needs. Although this may seem like a long shot today, scientists and entrepreneurs still believe they can someday develop safe, affordable nuclear fusion technology, which would instantly transform the entire debate about

energy. Current nuclear-energy plants work on fission, which is difficult to control and releases radioactive material. Fusion is the way the
sun generates energy—by fusing atoms together under enormous pressure. Whereas scientists have been able to set off fusion reactions, it's
always taken more energy to create the reaction than the fusion generates—not a great equation for supplying the world with energy.

Still, investor Peter Thiel funded a fusion startup called Helion Energy. The International Thermonuclear Experimental Reactor (ITER)
Fusion project is a joint effort by thirty-five countries working to prove
that fusion is feasible. The budget for this project is $20 billion, and it
aims to power up the reactor in 2025 for the first time. Tri Alpha Energy, funded by Microsoft cofounder Paul Allen and other investors,
has built a fusion machine that forms a ball of super-heated gas—at
about 10 million degrees Celsius—and holds it steady for five milliseconds without decaying away; five milliseconds is far longer than other
efforts have managed. The world invests almost $2 trillion in energy
every year, but just hundreds of millions of dollars go into fusion research and development, according to Tom Jarboe, an adjunct physics
professor at the University of Washington who studies controlled fusion. If we are dedicated to getting mankind off of carbon-based energy, pumping more money into fusion research would be a good
investment.

4

Healthcare

Genomics and AI Will Enable Prolonged Health

Most every week for a few years after joining Twitter, Othman Laraki and Elad Gil would get lunch, climb out onto the roof at the company's San Francisco headquarters, and eat and talk for an hour about technology and what they might do next. Laraki was the big-systems software guy with an MBA from MIT. Gil got his PhD in biology at MIT and was naturally interested in genetics. In the early 2000s they both worked at Google, and in 2007 the two cofounded Mixer Labs, which developed software that can help a cloud-based application learn more about its users' location. They sold Mixer to Twitter in 2009, hence the rooftop meetings.

During one of the meetings, in 2011, Gil brought along a hard drive. He'd paid about $5,000 to have his genome sequenced. A decade earlier that same process would have cost $1 billion. Now Gil's genome was right there on a simple drive. It was just data. This fascinated Laraki, whose family had a history of the BRCA mutation—the genetic predisposition for cancer made famous when Angelina Jolie in 2013 discovered she had it and opted to proactively get a double mastectomy to head off breast cancer. So Laraki asked whether he could borrow the drive and play with the data to see what he could find. "I told Elad that I'd use it to find all the bugs in him," Laraki jokes now.

Once Laraki dove in he found that the available software tools for analyzing genetic data were, to put it politely, poor. Or, as he more colorfully described them, "they stunk." "We were in the prebrowser days of genetics," he says—in other words, doing anything with genetic data proved as frustrating and balky as trying to use the internet before the arrival of the web browser in the mid-1990s. And in that, Laraki recognized an opportunity. Science had unlocked the ability to sequence the genome, extracting raw genetic data from our DNA, but there was still no good way to analyze and find meaning in the data in a cost-effective way. Solve that, Laraki and Gil believed, and it would become possible for everyone to affordably plumb their genetic data for secrets to living healthier, longer lives.

In 2013 the duo helped found Color Genomics, based on Laraki and Gil's insight about a broad swath of people being able to affordably obtain information about their genetic data. I invested, believing genomics will fundamentally change medicine. The team at Color Genomics built a software-driven service that could test for genes connected to cancer, such as BRCA. By automating the lab and genetic analysis with artificial intelligence and robotics, Color drove the cost of a genetic test down to $249—a price low enough to allow many consumers to afford the test without relying on insurance.

As I write this in 2017 Color and a few other companies such as Illumina are driving the costs of genetic testing ever lower while extracting yet more insights from the genetic data. The trajectory is clear: within ten years the cost will be so low that every child born can have his or her genome sequenced for current and future analysis (your genetic makeup doesn't change throughout your life). As costs go down the number of people who will have their genome sequenced will expand, including those now in their adult years.

As we gather this genetic data healthcare will be able to rely not just on doctors' instincts and experience but also on the hard data inside each of our bodies and the evidence it reveals. And genetic data is only part of this change. We're increasingly collecting other data about ourselves, such as vital readings from devices like a Fitbit, our health history in our electronic health records, and data from our mobile devices about where we've been and what we've been doing (e.g., if you have symptoms of Lyme disease, it's helpful during diagnosis to know whether you've been somewhere where ticks that bear Lyme disease are prevalent). All of this health-related data is becoming fodder for the leaders of startups who want to build new kinds of medical services that can cost-effectively understand what's going on inside each individual. Instead of treating diseases such as diabetes or high blood pressure based on what's worked for the broad population, you can treat your medical issues based on what will work for you. Drugs can be prescribed or even created based on what will be effective for you, even if they won't work for anyone else. That kind of research is going on in every major drug company lab.

This is the birth of personalized medicine—medicine for a market of one instead of mass-market medicine. In another decade the healthcare industry will look very different from how it looks today. Doctors will work hand in hand with artificial intelligence systems that plumb patients' data to understand a great deal about each patient's body. We won't need to build more mega-hospitals because unscaled businesses and services will be able to handle more and more of patients' needs. Drugs will be customized for each patient. Health professionals will know what treatments will work ahead of time instead of—as often happens now—waiting to see how a medication or procedure turns out. As unscaling plays out, healthcare has more potential to get cheaper, be more widely available, and be more effective.

* * *

It's instructive to understand why healthcare scaled so we can see how it can unscale. UnitedHealth Group has been a proxy for the healthcare industry's growth strategy over the past fifty years. The company was founded by Richard Burke, who grew up in Marietta, Georgia, and attended Georgia Tech in the 1960s, getting his BS in engineering, then an MBA and PhD. While in college Burke worked at an insurance company, processing health claims, and his interest in the field led him to a position at a healthcare think tank called InterStudy in Minneapolis, Minnesota.

By that time California pediatric neurologist Paul Ellwood, tapped to consult the Nixon administration about healthcare, was working on the concept of a health maintenance organization, or HMO, as a way to control healthcare costs. Until the 1960s healthcare in the United States remained something of a craft industry. Hospital chains were rare, and most doctors worked independently or in small groups. But by the 1970s, as the Baby Boom generation grew to adulthood and longer lifespans resulted in an increased number of elderly people, patients flooded the healthcare system, creating more demand than supply and driving up prices for consumers and for employers providing health insurance. HMOs were a potential remedy. They would offer patients or employers a subscription to healthcare as opposed to paying for every doctor visit and procedure, and in turn HMOs would open up a new way for healthcare businesses to pool resources, standardize healthcare for the mass market, and take advantage of economies of scale.

At InterStudy Burke developed new ideas for HMOs, and three years after joining the firm Burke wanted to build an HMO in the real world. So in 1974 Burke founded a small HMO, Charter Med in Minnetonka, Minnesota. It was soon reorganized under the name United-Health Group.

As UnitedHealth grew in Minnesota, it started buying other HMO groups and related companies to gain more economies of scale. In 1995 it paid $1.65 billion for MetraHealth, a group healthcare operation that was jointly owned by The Travelers and Metropolitan Life. In 1996 it bought HealthWise of America, which operated HMOs in the South. Then in 1998 it acquired HealthPartners of Arizona. It also bought health insurance companies and a Brazilian hospital chain. Essentially UnitedHealth went on a forty-year shopping spree that continued into the 2010s.

UnitedHealth proved to be the leading edge of a giant wave of mergers for scale across the industry. From 2007 to 2012, according to a Harvard report, 432 hospital merger deals were announced, involving 835 hospitals. Healthcare companies not only scaled horizontally but also vertically, pulling in every aspect of medicine. From 2004 to 2011 hospital ownership of physician practices increased from 24 percent of practices to 49 percent. By the mid-2010s 60 percent of hospitals also owned home health services, 37 percent owned skilled nursing facilities, 62 percent owned hospice services, and 15 percent provided assisted living options. UnitedHealth became the largest healthcare company in the world, treating nearly 40 million people a year in the United States and 5 million in Brazil. Other kinds of healthcare companies sought mega-scale too. Express Scripts grew to be the largest US pharmacy benefits manager, processing more than 1.3 billion claims a year. McKesson, a huge pharmaceutical distributor, now brings in more than $100 billion in annual revenue. Two companies, Labcorp and Quest Diagnostics, dominate the testing business. Johnson & Johnson, Pfizer, and a handful of other corporations have become goliaths of the global pharmaceutical business.

Scaling up healthcare over the past four or five decades was the right answer for the time. It had the effect of making good healthcare available to most of the population in advanced countries. It reflected a fundamental belief in much of the world that even the poorest people should have access to medical care. Outside the United States most developed countries implemented government-run universal healthcare programs to achieve that goal. Scaling up healthcare helped the general population get healthier and live longer—if nothing else, by simply supplying care to more people. In 1960 average life expectancy in the United States was 69.8 years. Thirty years later, in 1990, the average person lived 75.2 years. Today lifespans are around 78.8 years, according to the Centers for Disease Control and Prevention (CDC).

The economics of twentieth-century healthcare were built primarily on the concept of treating people *after* they got sick. After all, doctors and hospitals made more money in the fee-for-service model when people came in for more care. In more recent years the industry has pushed toward a fee-for-results model that incentivizes doctors and hospitals to keep people well. Yet there still has been a lack of data about individuals that could help foresee and prevent health problems, so the system has still had to rely on people realizing they were sick and then going to doctors to get treatment.

Of course, treating sick people is a lot more expensive than keeping people well, just as it's a lot more expensive to fix your car after it breaks down compared to the cost of regularly doing preventive maintenance. To profitably deliver healthcare to all sick people, the industry needed economies of scale to try to keep costs down and margins up. The more that medicine could be standardized, the better the industry could treat massive numbers of sick patients at a profit. That went for hospitals, doctor practices, and medical equipment makers. The most financially successful drugs have been those that work for the greatest number of sick people and for the most common conditions. Health insurance has worked best when it can spread the risks among the largest possible pool of customers.

But today scale is proving costly, and its benefits are stalling. In 2015 the United States spent $3.2 trillion on healthcare, or $10,000 per person. That was 17.5 percent of the whole country's GDP. "Of the $3.2 trillion health spending, 70 percent goes directly to fund the cost of our healthcare; the remaining 30 percent is spent on administration and profit, which is more than twice that of any other nation," wrote Vince Markovchick, president of the Healthcare for All Colorado Foundation, and Richard D. Lamm, former governor of Colorado, in 2016. "As much as $900 billion, or about one third of our total spending, can be attributed to waste, fraud and abuse."

Over the past twenty-five years healthcare inflation has been three to four times the rate of overall inflation. And even as healthcare costs shot higher in the United States, results for patients stopped improving at a relevant pace. When the World Health Organization ranked nations by healthcare quality, the United States landed at number thirty-seven, behind countries such as Costa Rica, Morocco, and Greece.

Gregory Curfman, editor of Harvard Health Publications, points out one way that scale has added to cost. "When individual hospitals merge into larger systems, they gain a larger share of the consumer health market," Curfman writes. "That puts them in a position to ask health insurance companies to pay more for medical care and procedures. These higher prices are not borne by the insurers, but by consumers in the form of greater premiums. Thus, some economists argue, mergers drive up healthcare costs and place added financial pressure on consumers."

Another problem with scaled healthcare: although the industry can scale up by building larger facilities and standardizing procedures, it can't scale up doctors the same way. The industry as it has worked in recent decades needs more and more doctors as the population expands and grows older, and that's not sustainable. One report by the World Health Organization says the world will be short of 12.9 million healthcare workers by 2035. If we continue to focus on treating sick people in mass-market healthcare, a shortage of doctors will mean more and more people will need to wait or travel to get treatment.

Scaled drug development, which brought us miracle cures ranging from Xanax to Lipitor to Viagra, is becoming ever more challenging. Drug development has been overtaken by what's called Eroom's Law—that's Moore's Law spelled backward because it is the inverse of the Moore's Law concept, which dictates that computing power constantly gets cheaper and better. Conversely, drugs constantly get more expensive and less impactful. A new drug now costs $2.6 billion to create, according to the Tufts Center for the Study of Drug Development. Pharmaceutical industry R&D costs increased a hundredfold from 1950 to 2010, even as labs implemented new technology to help the process. The pharma companies are challenged by what's known as the "better than the Beatles" problem: because most human conditions that can be treated with drugs are already being treated with drugs, any new drug needs to be much better than the old drug in order to reach mass-market success and bring in a return on the investment. This makes the discovery of new drugs much more difficult and expensive. Only a huge company can spend that much creating a drug, and it's only worth creating a drug if it's going to generate mass-market demand.

Unscaling and applying AI to massive new stores of data offers a way to lessen the need for big hospitals, ease demand for more doctors, and turn back the trend of out-of-control costs.

* * *

Glen Tullman, the CEO of Livongo, has a great way of describing healthcare's unscaling shift and sees it already happening—from personal experience. "Healthcare used to be the most local thing you did," he says. "You did it at home. And then, as we got more and more expertise, everybody had to go to Mecca—the big hospital." But the big hospital created its own set of problems. "If you put all the sick people in one place, you have secondary infection," Tullman notes. "The big hospital is hard to get to. It's inconvenient. So now we're starting to say, 'What if we went the other way? What if we could get doctors on demand instead of waiting for a doctor?' Initially we think that's for rich people. But not unlike Uber, if I told you some years ago, 'What I want is, I want a car, and at any time I want a car to pull up in front of my house and be waiting for me—take me anywhere I want it to take me,' you'd be, like, 'So you want a personal driver, a limousine and everything.' You'd say, 'You're crazy. You can't afford something like that.' Now everybody has that."

So why not believe everyone could have a doctor on demand? Why shouldn't healthcare unscale the way Uber unscaled on-demand transportation?

Tullman talks about his unscaled-healthcare experience. His son Sam, the football player mentioned earlier, broke his wrist two years in a row. Each time was a different wrist but the same exact break. Tullman took him to a big hospital the first time, where it took a full day to get treated, cost $5,000, and was a generally "terrible experience," he says.

When Sam's other wrist broke the following year, the surgeon who'd worked on the first one had opened a specialized, small surgical center in town. Tullman and his son drove up, parked in front, and were done in an hour and thirty minutes—for half the cost because the doctor's overhead and other costs were now so comparatively low. The surgeon had been able to build a center optimized for doing what he did. "He could actually get more done in a day because he didn't have all of the complexity of working in a hospital," Tullman says. "He could schedule more appointments. There was less chance of secondary infection with fewer patients in the building. Every aspect just became simpler, and patients were much happier."

Here's another example, this one from Long Island, outside of New York City, where a provider called Northwell Health launched a service called House Calls, which is aimed at elder care. Geriatricians have warned for years that hospitalization can accelerate an elderly patient's decline. Yet when an older person falls or has chest pain, an ambulance usually delivers the patient to a hospital. House Calls is an attempt to keep elderly patients out of the hospital. The paramedic arrives in an SUV, not an ambulance, with medical equipment that can send information back to a physician on call. "A lot of what's been done in the E.R. can safely and effectively be done in the home," Karen Abrashkin, an internist with the House Calls program, told the *New York Times*. For frail, older people with many health problems, Abrashkin noted, "the hospital is not always the safest or best place to be." In fact, as the *Times* reported, Northwell Health's community paramedics program published its results in the *Journal of the American Geriatrics Society*, looking at outcomes for 1,602 ailing, homebound patients (median age eighty-three) over sixteen months. When the community paramedics responded—most commonly for shortness of breath, neurological and psychiatric complaints, cardiac and blood pressure problems, or weakness—they were able to evaluate and treat 78 percent of patients at home. By relying on portable technology, mobile communications and the data that can be sent back to the hospital and instantly analyzed, Northwell Health has been able to peel off a slice of hospitals' customers and serve them better—the very nature of unscaling. In the same manner the economies of unscale are increasingly supporting smaller, more focused healthcare operations that profitably unbundle a segment of demand—broken wrists, geriatric care—and serve that segment better and at lower costs to the system.

Instead of treating every person with diabetes the same, an unscaled, data-driven healthcare system will treat each individual's diabetes. In coming years, instead of giving you a pill that millions of others take and hoping it will work, you'll get a pill that might be tweaked just for your genetic makeup so the doctor knows it will work (and won't give you a horrible side effect) before you take it. Instead of going to the doctor after you get sick, your "doc in the cloud" will constantly monitor data about you and know when your patterns are changing in a way that suggests something is wrong—and inform you to take action or see your doctor before you get sick.

In a world of unscaled healthcare the best business opportunities will lie in keeping people healthy so they can avoid doctors or hospitals as well as the costs that go along with them.

* * *

The driving force behind unscaling healthcare is the data about our health that is now pouring in and the artificial intelligence software that will help make sense of the data. Genomics will exert enormous force on healthcare, and the upside is almost impossible to describe today. A parallel is the time during the 1970s when information was just beginning to be digitized. At that time it would have been hard to envision today's databases and digital media and data analytics. We're at such an early point in genomics; not even .01 percent of the population has had its genome sequenced, according to UBS Securities. The cost of getting genetic information is falling faster than Moore's Law would dictate, diving from a project only a government could fund to, well, Color's $249 test. It's not hard to see that the cost will continue to fall until it's the same as getting a routine blood test. The Broad Institute, an MIT and Harvard genomics research center, predicts that the boom in sequencing will collect a zettabyte (1 sextillion bytes) of data per year by 2025. A zettabyte is equivalent to all the internet traffic on the planet in 2016.

As the price of genetic testing drops, more and more people are becoming increasingly aware of different reasons to send a DNA swab to a lab. For instance, 23andMe, which sells its service directly to consumers, offers a genetic test that includes reports about ancestry, personal traits such as balding, and whether you're a carrier of genetic conditions like cystic fibrosis. Ancestry.com lets consumers mail in a saliva sample and sends back a report on ethnic origins. Ancestry claims to have the world's largest consumer DNA database, with samples from more than 4 million people. All this consumer activity is more than just fun—the more of the population that gets tested, for any reason, the more data we'll gather on genetics. More data means we'll learn more and make genetic testing even more valuable, helping us drive toward a time when everyone benefits from getting their DNA sequenced.

The data from genetic testing will be key to personalizing medicine. "We can use new innovations in genetic testing to more precisely predict, diagnose, and treat conditions in individual patients, instead of having to settle for a one-size-fits-all approach," explains Dr. Andrew Gurman, president of the American Medical Association. Researchers can use genomics to target and treat rare genetic disorders that, according to the CDC, affect about 25 million people in the United States. The information will lead to many innovations in healthcare, such as personalized cancer vaccines. Dr. Pramod Srivastava, interim chairman of the Department of Immunology at the University of Connecticut, as of this writing was recruiting patients for trials of the world's first ovarian cancer vaccine based on genetics. "Previous personalized cancer vaccines acted on faith; with genomics, we can actually know how each patient's vaccine is unique. It is a remarkable transformation of the whole field of cancer immunology and immunotherapy."

Genome data is revelatory about the human body, but it's also limited to what happens inside your cells. In this new era of medicine doctors will be able to cross-reference DNA data with a flood of additional data about many other important elements that affect your health, like your vital signs, activity, what you eat, where you travel. All of that will be gathered by connected devices—the Internet of Things.

Tullman's Livongo is one of those companies bringing IoT to healthcare. Livongo's connected devices take blood glucose readings for people with diabetes, sending that information back to an AI-based system. Devices will increasingly be able to take any kind of reading imaginable and send it back for analysis—heart rate, blood pressure, temperature, hours slept. One company, Future Path Medical, makes a device called UroSense that can monitor urine in catheterized patients, and the data can help foresee kidney problems and prostate tumors. Doctors will order some devices; many will be bought by consumers who want to learn how to keep their bodies healthier and performing better. Analyst group Research Beam forecast the IoT healthcare market to hit $136 billion by 2021, up from $60 billion in 2014.

Meanwhile our cell phones know where we are and where we've been, which might help diagnose a disease or flu that's common to a certain part of the world by cross-referencing where a person's been to known contagions in that place. As we order food and groceries online, the data about what we eat—a huge factor in our health—can merge with other medical data to help a diagnosis. Our online activity can track how much we work and exercise. All this data can be revelatory about our health. At some point health data might all flow into some kind of personal health app that connects with all sorts of other apps and devices. "Facebook has a lot of valuable personal data, but much of this value comes from giving other applications on your phone access to your Facebook profile—you can log into many iPhone apps using your Facebook profile," notes a report on new health technology by UBS Securities. "This increases the value of the Facebook data. We could envision a similar dynamic where a consumer genomics company could increase its value by becoming the hub of personal health data, for example feeding into tools like Apple Health, or tying into lifestyle tools like Fitbit."

To all this new data we'll add one more important information stream: our personal health records. Electronic health records (EHRs) are finally taking off after years of discussion. In 2009 16 percent of US hospitals were using EHRs; by 2013 that had rocketed to 80 percent, according to *Becker's Hospital Review*. As our health records become digital and searchable, patients can get access and take more control of their healthcare. The data from IoT devices, genetic testing, lab tests, and doctors' notes can all combine to generate deep knowledge about each individual's body and health, teeing up the ability to treat each patient as a market of one instead of as simply a part of a mass market.

This is where artificial intelligence plays a crucial role. Because AI is software that can learn, an AI system can take in data about you, get to know you, and identify patterns in your health. As more information comes in—say, from IoT devices monitoring your heart rate or blood glucose—AI software can look for changes that suggest you might be getting sick. The system will catch illnesses early, when they're easier to treat.

If you have a medical problem, you probably won't go to a hospital that has a wide range of services because you'll already know what's wrong and can seek out a small, specialized way to treat it. Mass demand for all-inclusive hospitals will give way to individuals demanding focused healthcare—maybe from a corner clinic that only handles the specific condition you need to get treated.

* * *

Going to a doctor's office and waiting in a fluorescent-bathed room full of sick people has been a common experience ever since doctors ended house calls. Yet as a sign of what's coming, a startup called Spruce—founded by Ray Bradford, who'd been an executive at AWS—is a step toward allowing you to "see" a doctor through your smartphone. Spruce started with dermatology. Download the app, and it offers a choice of conditions: acne, eczema, bug bite, and so on. Based on the condition, it asks a series of diagnostic questions ("How would you describe your skin? Normal? Oily? Dry?"). Then it tells you what kinds of pictures to take of your condition. Finally, you can pick a participating doctor to send everything to or just choose "first available." You're promised a response within twenty-four hours, most of the time with a prescription sent to your drug store and instructions telling you how to care for your condition. The cost is $40—not much more than the copay for many specialist visits, which means using Spruce can make sense even if it's not covered by insurance.

Spruce is one startup in a bubbling stew of activity around handheld medicine. You can't even call it "telemedicine" anymore—that term has been around a while and usually means real-time, Skype-like video interaction. The new breed is rethinking medicine based on smartphones, mobility, and cloud-based AI. Handheld medicine companies Doctor on Demand and HealthTap have both referred to themselves as an Uber of doctors, and they've each raised more than $20 million in funding. Other apps can take blood pressure and electrocardiogram readings and send them to a cardiologist. Apps for eye tests are becoming as accurate as those steam-punkish contraptions ophthalmologists push up to your face—it probably won't take Warby Parker long to offer a prescription eye exam on your phone.

As doctors increasingly work with data and AI, they will be able to diagnose patients much more quickly and accurately than today, and this has the potential to drastically lower healthcare costs and make people healthier. IBM's Watson AI technology, already being tested at Cleveland Clinic and a handful of other health institutions, is showing how it can work alongside a doctor, asking a patient questions to help arrive at an accurate diagnosis. The Watson systems are fed millions of words of medical research and case studies—far more than any doctor could ever read. (Medical information doubles every five years; 81 percent of doctors say they spend five and a half hours or less reading medical journals each month. Doctors can't possibly know any more than a fraction of what's in the latest literature.) When a patient comes in with an unusual condition, a doctor and Watson can interact with the data to narrow down possible conclusions. Not only does that help patients get better treatment; it also helps doctors come to a diagnosis faster, with less time spent on research. "We feel like it has a strong potential to address the problem of physician burnout and the challenge of being mired in data and not actually having synthesized knowledge," says William Morris, the Cleveland Clinic's associate chief information officer.

Over time AI systems such as Watson will be able to get to know patients while constantly reading and learning more about medicine, leading to better and better answers for the doctors who consult with it. Anyone running a modern hospital will be charged with figuring out how to make data course through the hallways and operating rooms and how to turn that data into better care.

"As a computer science guy, I see that the way doctors operate today is like expert-systems software," Color's Laraki says. "They rely on a decision tree that we made simple enough to fit into an average smart person's brain. We load it up in medical school. But later its decisions are based on a limited set of inputs and complexity." The opportunity and upside now is that the doctor, working hand in hand with AI-driven systems, will be able to access a level of data and medical knowledge that would never fit inside one person's head. "The doctor becomes less like someone using a rare superpower and more like a data practitioner," Laraki says.

Data and AI together can help robotics "learn" to do a job better. In healthcare, robots are already helping with delicate procedures like eye surgery. There is little doubt that robots will eventually be better than humans at some kinds of surgery. They are more precise than people and never have an off day. Human doctors won't be eliminated from the operating room; instead, they will become collaborators with robots, sometimes remotely. And if robots can be programmed to mimic the best surgeons, then the best surgeons can essentially be made available in many small clinics instead of a few huge hospitals. More than two hundred companies are already active in various aspects of the healthcare robotics market, creating highly specialized devices for a wide range of applications. The da Vinci Surgical Systems robot, approved by the US Food and Drug Administration (FDA) in 2000 to assist surgeons in minimally invasive procedures, is so precise that it can peel a grape. Yet some of these robots will take time to perfect and become accepted. For instance, Sedasys is a robot developed by Johnson & Johnson that can deliver and monitor anesthesia for short surgeries without an anesthesiologist present. The FDA approved it in 2013 after it underwent extensive safety trials. And it could drastically lower costs—the charge for using Sedasys for anesthesia was about $200, compared to $2,000 for a human anesthesiologist. Yet, as you might imagine, the human anesthesiologists complained vociferously. Hospitals then stopped buying the robots. In 2016 J&J reluctantly stopped making them.

Most mass-market pharmaceuticals are the epitome of imprecision medicine. The top-ten highest-grossing drugs in the United States help between one in twenty-five and one in four of the people who take them, according to *Nature*. For some drugs, such as statins—routinely used to lower cholesterol—as few as one in fifty may benefit. Think about it this way: to get US FDA approval, a drug company needs to prove that a drug is safe and works for most people. But *you* are not most people. You are you. Let's say a cancer drug would make half the population worse but save the lives of the other half. It would not get approved. But if data plus AI—personalized medicine—could absolutely know before you take the drug that your genetic makeup means you'll be one of the patients whose life it saves, then the whole drug discovery and approval concept becomes a different game. "It may seem that an N=1 experiment (a drug trial that only applies to one person) is not scientifically valid, but it turns out that it is extremely valid to you," writes Kevin Kelly in *The Inevitable: Understanding the 12 Technological Forces That Will Shape Our Future.* "In many ways it is the ideal experiment because you are testing the variable X against the very particular subject that is your body and mind at one point in time. Who cares whether the treatment works on anyone else? What you want to know is: How does it affect me? An N=1 provides that laser-focused result." As you can imagine, the N=1 approach would give the FDA fits. The agency is set up to test drugs on the general population. The FDA's entire process would need to be overhauled.

Still, if researchers can get to the point where data can tell us the effect a drug will have before the patient takes it, then more drugs can be approved, driving down the cost of drug development. A new drug would no longer have such a burden of proof, and if it costs less to develop, it could be profitable even if it has only a small potential market. Small entrepreneurial drug makers will have an opening to rapidly develop new drugs for niche markets, successfully competing against Big Pharma. Ultimately the healthcare industry's expectation is that data plus AI will mean that drugs can be customized for any one person. Genetic data will help a doctor figure out the compound that will work on a particular patient, and then a drug company will be able to make the one-off drug for that person instead of manufacturing and distributing giant batches of drugs.

The coming flood of genetic and health data will have an enormous impact on insurance as well. Political debates in the United States over universal healthcare are, in a way, the wrong debate. Policymakers should be working on ways to make insurance available to a market of one—that is, to you. As you might imagine, this can get tricky. All that data that will help doctors predict when you'll get sick can also predict your future health for an insurance company. Today insurance works because the premiums from relatively well people subsidize the high expense of covering sick people. But if an insurance company can accurately predict how much it will cost to take care of you, it could charge you premiums based on that amount. That should mean that people who take care of themselves and have little genetic predisposition for disease should pay very low premiums based specifically on them. People who smoke or don't exercise or whose genes predict a future health problem would need to pay more. That could come to seem grossly unfair, with some people paying far less for health insurance than others. How to manage that should be the debate among policymakers.

But there's another dynamic that should take hold with insurance. Personalized, predictive, unscaled medicine should keep more people healthier for less cost—because it will catch problems like cancer or heart trouble very early, when it's easy and cheap to treat. The more of your health data that you allow the insurance company to access, the cheaper your rates should be—because with more data, health professionals will more likely stop diseases early. Yes, it's a major privacy risk, but we'll each need to weigh that risk against how much we want to pay for healthcare and insurance. People who don't want to give up their data will wind up paying a great deal more—even more than sick people who do open up their data. That will be the trade-off we'll each have to make.

* * *

Healthcare is shifting from analog to digital. Though much more complex, the shift is not all that different from when, for instance, music moved from analog (vinyl records, cassette tapes) to digital (CDs, downloads, and now streaming). Medicine, for so long centered around the knowledge in the human brains of doctors, radiologists, and other experts, is turning into a data science problem. The whole focus of healthcare is shifting from treating conditions after they arise to keeping people healthy and stopping illnesses before symptoms appear. And it is unscaling, moving from a mass-market scaled-up approach to an unscaled market-of-one approach. All this opens up enormous opportunities for technology to recast entrenched practices. Back in the mid-2000s I rarely even considered healthcare startups. Today it's one of the most exciting areas of investment for our firm, and we're seeing a constant stream of innovative new companies taking healthcare products to market.

Here are some of the opportunities I see coming:

PERSONALIZED HEALTH TECH: To help people with diabetes manage their disease, Livongo built a wireless device that's as easy to use as a smartphone, and it connects through the cloud to Livongo's software and to human-driven services if necessary. The market is some 30 million people with diabetes in the United States alone, not to mention hundreds of millions more around the world. And that's just one disease.

There's no question we're going to see a boom in connected medical devices and services designed to help people better understand their health or deal with conditions ranging from high blood pressure to cancer. In September 2016 CB Insights, which tracks venture-funded technology companies, identified seventy-two connected medical device startups. To give an idea of the range of devices coming, one company, OrthoSensor, makes a device that goes into a knee replacement and lets doctors know whether the new knee is performing correctly; a company called Vital Connect makes HealthPatch, which looks like a big bandage and can read a person's heart rate, respiratory rate, skin temperature, and body posture and then send it back to a doctor or hospital; a German company, VivoSensMedical, is marketing the OvulaRing, which is inserted into a woman's vagina and transmits back information that lets her know when she's most likely to conceive.

Another twist on medical devices will be connected drugs, intended to help patients take medication the way they should. (The healthcare industry believes about half of medications aren't taken as prescribed.) For all of history pills have been dumb—the drugs don't know anything about you or whether they're working the way they should. When you fill a prescription in the coming years, you'll get a bottle of pills plus software that analyzes biological data from your phone, wrist monitor, networked bathroom scale, and so on to figure out if the drug is working and whether your doctor needs to modify the dosage. It will funnel all that data into an app intended to engage you in your treatment—because if you see progress, you're more likely to keep taking the drug. Major pharmaceutical companies are beginning to think about digitizing drugs. Merck founded a unit called Vree Health a couple of years ago to explore "technology-enabled services."

If you put all of that sensing technology together, it's going to become possible to create a virtual hospital. Instead of working in a scaled-up single building as a hospital, doctors will be able to monitor patients wherever those patients happen to be—at home, in small clinics. The hospital will be in the cloud, tying together the sensors and doctors and patient management through networks and software.

PERSONAL HEALTH DATA: Although the healthcare industry has made progress digitizing patients' health records, it's been a slow and clunky process. The records are not standardized and often not easily shared with patients, doctors, or medical researchers, much less across technology systems. As data becomes a more important part of good healthcare, patients are going to want more access to and control over their information. It's a huge opportunity for startups to attack in new ways.

Our firm has funded one, called PatientBank. As cofounder Paul Fletcher-Hill has said, the company started when he and the other founders decided to build a medical-record app using their own information. "We assumed it'd be easy to access our data but quickly realized that was not the case," Fletcher-Hill said. "To get our records we had to mail in formal requests to all the hospitals we'd ever been to, and we ended up with piles of paper to sort through. We decided there had to be a better way." Now the company's app gathers medical records on behalf of individuals and businesses. Rather than faxing in documents or visiting the hospital to make an information request, PatientBank lets anyone request their files online and receive it in about ten days—three times faster than the average request.

We need to develop what I call the "personal health cloud"—where all your data gets collected and is easily accessible by people or software with permission. Many companies are working on it, including KenSci in Seattle and iCare in San Francisco, and I've been helping a new one get off the ground. The industry needs an open platform on which innovators and entrepreneurs can build applications and services, much as they can build on Apple's app store.

MEDICAL AI: Artificial intelligence will be the force that holds together the coming onslaught of medical data. It will be key to niche services, much as it drives the back end for Livongo by learning about each user's readings and spotting an unusual spike or drop. It will also be deployed in general large-scale ways, helping doctors and hospitals track and understand what's going on with their patients.

IBM is making a big bet on medical AI. "I believe something like Watson will soon be the world's best diagnostician—whether machine or human," says Alan Greene, chief medical officer of Scanadu, a startup that is building a diagnostic device inspired by the Star Trek medical tricorder and powered by a medical AI. "At the rate AI technology is improving, a kid born today will rarely need to see a doctor to get a diagnosis by the time they are an adult."

Others are working on more focused applications of AI for medicine. A group of Stanford computer scientists, with the help of well-known AI and robot pioneer Sebastian Thrun, developed AI-driven software that can diagnose skin cancer as well as dermatologists. The research team ran a database of nearly 130,000 skin disease images through their AI, training the software to visually diagnose potential cancer. "We realized it was feasible, not just to do something well, but as well as a human dermatologist," said Thrun. "That's when our thinking changed. That's when we said, 'Look, this is not just a class project for students, this is an opportunity to do something great for humanity.'"

AI will infiltrate every area of healthcare. It will help read results from X-rays, MRIs, and other tests, often seeing patterns humans wouldn't. AI will help drug companies learn what compounds work for what kind of genetic makeup. It will help health officials spot broad trends in consumer behavior so as to spot disease outbreaks early and try to contain them. In fact, I would say that if a health startup came to us and was *not* using AI, I would probably not be interested.

NEW WAYS TO BE A DOCTOR: Some doctors already get how much the profession is going to change. "I would like to eliminate the waiting room and everything it represents," David Feinberg, a doctor and CEO of Geisinger Health System in Danville, Pennsylvania, told *Becker's Hospital Review*. "A waiting room means we're provider-centered—it means the doctor is the most important person and everyone is on [his or her] time. We build up inventory for that doctor—that is, the patients in the waiting room. We need to increase access and availability so we can show people we see it is a privilege to take care of them—to tell patients, 'We are waiting for you.'"

At a time when data and apps can do a lot of the diagnostic and preventive work of many kinds of doctors, the doctor will need to become less of a mass-market processor of patients and more of a service provider focused on the individual. The very idea of an annual physical will become radically different. We will all get our genome sequenced, and that data will sit in the cloud along with all our medical records and other data about our health and lifestyle. A physical should then be more like getting your car inspected—plug into the computer, analyze the readings, and you get a view of the patient's health that's better than anything doctors do in their offices today.

All this opens opportunities for new types of data- and patient-centric clinics and doctors' groups that can peel off aspects of healthcare now bundled inside hospitals and big mass-market health groups.

HEALTH INSURANCE: Healthcare is an oddity in business—it's perhaps the only sector in which insurance is actually part of the ecosystem. As healthcare changes—data driven, patient centric, unscaled from big hospitals—insurance will surely change. The biggest opportunities will come once cloud-based EHRs can capture all our health data in a way that allows us to control what we share with an insurer. New kinds of insurers will use this data to offer health insurance customized specifically to you and your risks instead of lumping you in with everyone who is broadly similar to you. The insurer can offer a bargain: the more data you allow it to view, the better financial incentives it can offer for you to stay provably (via the data) healthy. This way individuals will be more responsible for their insurance costs based on their actions.

GENOMICS: We're at the beginning of what UBS Securities calls "the genomics big bang." The sector is about to explode. The cost of sequencing people's genomes is falling faster than the Moore's Law curve, although only .01 percent of the population has so far had their genome sequenced. Consumer genetics—driven by companies like Color, Ancestry.com, and 23andMe, which sell directly to consumers and give customers access to the results—is predicted to take off the way cell phones did in the 2000s. Genetic data will become the most important fuel powering the unscaling of the healthcare industry.

Genomics startups will offer genetic screening for predisposition to diseases such as cancer or Alzheimer's and physical traits from baldness to obesity; genetic matching to determine drugs that are more likely to be effective specifically for you; and genetic data analysis that can be used to guide medical treatments for the rest of your life.

In another decade science is likely to unlock a whole new level of genomics: gene editing. A breakthrough method called CRISPR-Cas9 came to the public's attention in 2013 when researchers first used it to precisely slice the genome in human cells. It immediately became controversial because the technique can edit what's called germline cells, which get passed on from generation to generation. It seems to hold the possibility of editing genes to create smarter, stronger, more attractive people who will reproduce from generation to generation, perhaps giving rise to a massive human divide between the perfected people and the rest of the population. For that reason editing germline cells is illegal in some countries and hotly debated in others.

But CRISPR-Cas9 can also be used to edit somatic cells, which don't get passed on but can be responsible for some genetic diseases or defects. Companies and labs are racing ahead to develop safe and effective ways to do this. In late 2016 Chinese scientists for the first time injected a person with cells modified using CRISPR-Cas9, in a test that attempted to treat aggressive lung cancer. The test is driving yet a faster pace of scientific work on gene editing. "I think this is going to trigger 'Sputnik 2.0,' a biomedical duel on progress between China and the United States, which is important since competition usually improves the end product," Carl June, who specializes in immunotherapy at the University of Pennsylvania, told *Nature*.

So scientists are still years away from making gene editing safe and effective enough for routine use in healthcare. But once that day comes, I'll be looking for the next great gene-editing business idea.

5

Education

Lifelong Learning for Dynamic and Passionate Work

Sometimes it's helpful to come from outside the system you want to change. Jeff Bezos wasn't in retail before he started Amazon.com. Patrick and John Collison weren't in finance before starting Stripe. And Sam Chaudhary wasn't an educator before he built ClassDojo and began trying to reshape the classroom. Chaudhary grew up in a small seaside village in Wales, and while he was in grade school his family moved to Abu Dhabi. The international school he attended there often asked the better students—and Chaudhary was always one of the better students—to help teach the others. So, as Chaudhary says, "I ended up teaching for something like twenty hours a week from when I was twelve to when I was eighteen." Looking back now he believes that made him want to work—somehow, in some way—in education.

For a while, though, the education industry was not where Chaudhary was heading. He went to England's Cambridge University to become an economist, planning to get his PhD. Big banks were recruiting him, but finance didn't appeal to him. Instead, he briefly taught at a high school. Consulting giant McKinsey, which had also been trying to recruit Chaudhary, heard about his teaching stint and convinced him to join the firm to work on education projects. Chaudhary did that for a couple of years—until he met Liam Don.

Don was born in Germany, grew up in London, had a computer science degree, and was a game developer. "We just got on and decided we should work together," Chaudhary says. "We wanted to work on an education thing together." They only knew that cool game technology might be a useful way to help kids learn in classrooms. With no solid idea about what they were going to do, they moved to the San Francisco Bay Area and applied to work in a small technology company. "We hadn't lived or worked here [in the United States] before," Chaudhary says. "We didn't have experiences in schools here. So the natural thing was to talk to teachers." They cold-called thousands and spoke to hundreds, trying to get to the heart of the classroom challenges that teachers faced, looking for a way to perhaps overcome them by applying new technology.

As Chaudhary and Don listened to teachers' frustrations—often around needing to spend too much time managing behavior in the classroom and too little helping kids learn and advance—the two started to understand why there is widespread agreement that education needs to change and yet little change occurs. For about 150 years society has been scaling up schools and modeling them on factories and corporations. Such schooling was the right answer a century ago, once the Industrial Age superseded agriculture as the basis for progress and prosperity. Education should prepare students for the world they're going to enter, and the factory model of schools—complete with standardized ways of doing things, bells to signal the start and end of classes, a clear hierarchy that separates students (like labor) from teachers (i.e., managers)—certainly did that for a long time. But in this century the Industrial Age has been declining as the Digital Age ascends. "Yet teachers are expected to operate the same way they would've fifty or a hundred years ago," Chaudhary says. Reformers mistakenly approached change in education as if they were redesigning a factory—from the top down. "Education is not a mechanical system," Chaudhary observes. "It's a human system. It has to be changed from the bottom up." Real change, he concluded, needs to start with teachers, students, and parents, one classroom at a time.

The interviews with teachers that Chaudhary and Don conducted, plus how they thought about the history of schools, led them to a core problem they believed they could solve: teachers needed a way to improve student behavior and to create a classroom culture more like that of a high-performance team—in fact, like a team that might build a startup or collaborate on a project. That would more closely match work in today's world. That insight led Chaudhary and Don to build a classroom communications app—a way for teachers to give kids feedback, loop in parents, post photos and videos from the school day, and create a digital community to complement the real-world classroom. They called it ClassDojo, and it allows a teacher to create a mini social network that includes the teacher, the students, and the parents or caretakers. The teacher can post videos or announcements, update individual parents on their child's work and behavior, get feedback from parents, and generally create an ongoing communication loop for the class. Better-informed parents tend to stay more involved with their kids' schooling, which helps the students and boosts the chance that kids will do their homework and behave well in class, which in turn frees teachers to do less policing and more teaching. "It's basically an app where you can get rewarded for positive behavior," said Morgan Costa, a fourth-grade student at Highlands Elementary School in Danvers, Massachusetts. "It's really fun because at the end of the week, whoever gets the most points, gets to be the teacher's assistant, and we get rewarded with good expectations. And our parents can go through the app, and log in, and see how we are doing."

The startup launched its first version of ClassDojo in August 2011. After five weeks, more than twelve thousand teachers had signed up, all through word of mouth. By ten weeks the number had grown to thirty-five thousand—about 1 percent of all teachers in the United States. (This kind of uptake helped convince me to invest in their company.) By mid-2016 ClassDojo was being used in two-thirds of US schools and had spread to 180 countries. Teachers were sharing moments throughout the day with parents and making parents part of the classroom culture. Judging from the product's rapid spread with almost no marketing, teachers seem to feel that it makes their classrooms—and their working lives—much better.

On the surface ClassDojo looks like just a useful app for teachers, but it's one sign of the larger forces that will unscale learning. Look at it this way: a big school or school district is an integrated bundle of classrooms, like a big corporation. In that model the teacher is a middle manager, doing what the higher-ups dictate. Parents often aren't even in the equation.

ClassDojo is a rentable platform—free for teachers—that helps a teacher create a culture around her classroom community and more easily tailor education to each student. The community that the teacher, students, and parents experience revolves thus more around the classroom—tied together by ClassDojo—than around the school as a whole. In other words, the app helps make "school" seem smaller, more personal, more customized than when "school" is a big and often bureaucratic bundle of classes of many different age groups in a large building.

ClassDojo has also been adding interactive lessons to its app. For example, in mid-2017 the company formed a partnership with the Yale Center for Emotional Intelligence to bring short videos of mindfulness lessons into the app. Other content might help kids with math or history. Students can engage in the lessons at their own pace, whenever they want. In this way, more learning happens in the app, and the teacher can become a coach and guide for each student, not just someone who repeats a state-dictated lesson plan.

As young students grow up using technology like ClassDojo to obtain a more customized and intimate education, they will come to expect that kind of learning later—at college, at work, and throughout life. It is just the beginning of reshaping learning for the new era.

* * *

Horace Mann could be considered the father of scaled education in the United States. Mann was born in Massachusetts in 1796, just twenty years after the American Declaration of Independence. Agriculture dominated the global economy, yet the Industrial Revolution had begun in England, where James Watt was perfecting his steam engine. In 1816 Mann, at the age of twenty, was accepted into Brown University in Rhode Island and studied law. Eleven years later he won a seat in the Massachusetts state legislature, where his career in politics led him, in 1837, to help create and run the state's first board of education.

By then Mann had traveled in Europe, where he saw factories increasingly driving the economy and new kinds of schools that had been set up with the factory model in mind. "Employers in industry saw schooling as a way to create better workers," writes Peter Gray, a Boston College historian of education. "To them, the most crucial lessons were punctuality, following directions, tolerance for long hours of tedious work, and a minimal ability to read and write." You can see in that statement a reflection of schooling even in the twenty-first century. "The idea began to spread that childhood should be a time for learning, and schools for children were developed as places of learning," Gray concludes.

Before Mann's time most children in the United States were educated—if at all—in one-room schoolhouses. All ages shared the same classroom, and teachers had no common curricula. Mann, inspired by the European factory-school model, set out to change all that. His education board introduced standardized curricula to be taught throughout the state and implemented what was then called age grading, which put students of the same age together in a "grade" and gave them the goal of graduating to the next grade. Students, in a sense, were being processed through an assembly line, like a product in a factory: enter at age five or six, go through standardized lessons taught in the same precise way by every teacher, and leave the factory school at age eighteen as a completed product. Mann's innovations were so successful that one state after another adopted them.

Once schools were standardized, they could be scaled. Putting many classrooms together in one building achieved economies of scale, serving more students at lower costs. As the population ballooned, the same standardized education could be replicated to serve it. Schools got bigger. By the 2010s some of the largest high schools in America—Brooklyn Technical High School in New York; Reading High School in Reading, Pennsylvania; Skyline High School in Dallas, Texas—had from forty-seven hundred to eight thousand students.

Universities followed a similar path. In Mann's years at Brown, college was a craft industry. A relatively small number of distinctive small campuses around the world offered higher education to a tiny sliver of the population. Then, in 1862, the US Congress passed the Land Grant College Act, which helped establish a new era of large state universities open to a greater number of the public. College, in effect, could become a factory too. Put students in after high school, have them all go through a largely uniform four-year process, and they'd come out the other end ready to be professionals. As I write this, the University of Central Florida, the largest college in the United States, has sixty-three thousand students. Ohio State, right behind it, has nearly fifty-nine thousand.

Scaled education has been one of the great drivers of progress the world over. It helped create generations of factory workers, managers, entrepreneurs, innovators, scientists, politicians, writers, and artists who, in turn, built businesses and institutions that brought us prosperity. Scaled education lifted vast numbers of people out of poverty and into a broadening middle class. In the mid-2010s about one-third of the US population held a four-year college degree—a remarkable achievement.

As well as scaled education worked, however, it was built for an age of scaled business and institutions. It is a mass-market, one-size-fits-all model of learning. It doesn't do a great job of preparing students for the unscaled, entrepreneurial economy we're entering now. It doesn't take advantage of new technologies to help students learn in new ways that cater to who they are as individuals.

In higher education the scaled model has become unsustainable. The cost of going to college for a four-year degree has been inflating by more than 5 percent a year for almost two decades. By one measure, if this trajectory continues, private college for a child born in 2010 will end up costing almost $350,000. In the mid-2010s a Kauffman Foundation study found that the rise in student debt in recent years coincided with a decline in new business startups. One interpretation of this is that young people are so burdened with debt that they can't take a chance on entrepreneurship—they have little choice but to take a wage-earning job to pay off the loans. That means that the cost of scaled higher education is hurting society in important ways.

For decades US politicians, educators, and the public have been concerned that our schools aren't properly preparing students for our world. They keep trying to reform education, but almost every attempt completely misses the problem: that school as it is today was actually built to serve a different world—a previous economic era that's now waning. The reformers try to redesign, from the top down, our factory-based scaled education model by setting up, well, some other factory-based scaled education model. But building a new scaled education system—constructing new schools, buying new equipment and books, hiring scores of new teachers—can be expensive and slow.

Instead, we need an unscaled approach that will better fit an unscaled economy. Curricula need to change from standardized factories into personalized processes for a market of one—because that's how the new economy works. Education needs to take advantage of the new technology wave and reinvent how everyone learns—not just children but also midcareer adults. For ages we've made people learn in ways that aren't natural so we could do it with a standardized, scaled-up system. We made people conform to the system rather than the other way around. In an unscaled world the system will conform to people. Technology will enable you to learn in a way that's tailored to you.

* * *

Khan Academy has been one of the most impactful experiments in education in the AI century. Its founder, Sal Khan, is my old friend from MIT, and I'm on Khan's board. Starting with rudimentary YouTube videos in 2006, the not-for-profit Khan Academy developed into a sophisticated, AI-driven online platform for teaching each student at their own pace. Microsoft cofounder Bill Gates, through his Bill and Melinda Gates Foundation, has donated more than $9 million to Khan Academy, and Mexican billionaire Carlos Slim donated millions to help Khan expand Spanish-language courses. By 2017 40 million students and 2 million teachers used Khan Academy every month, in thirty-six languages. Courses in math, for instance, range from simple counting for kindergartners to differential calculus for college students. But as much as Khan Academy has affected learning for a slice of the global student population, Sal Khan has bigger ambitions.

"We imagine that in ten years you get access to a low-cost smart-phone or a computer, and you can self-educate from the basics like your letters and numbers all the way through vocational skills or skills that allow you to plug into formal academic systems, go to college, become a professional." In other words, Khan wants to make it possible for a student to use Khan Academy to educate herself without ever stepping into a school building and to wind up qualified for the top jobs anywhere.

He's not intending for classrooms to disappear, although children in regions where there are no classrooms nearby would still be able to learn just by tapping into AI-driven, cloud-based coursework. "The physical environment is still going to play a very important role," Khan says. "That's where you're going to get a lot of social and emotional development and your work habits. School won't be about grades—it should be more about the kind of community you build, the social interaction. So school will be a place where children gather to learn how to work in teams to get things done, how to form a community and socially interact, and perhaps, most importantly, school will be a place where you learn how to learn and what to learn—so you can then go online to get the actual content and information that you need to know. This style of learning fits more with the technology and econ-omy we're creating today.

A classroom will become a community built around the teacher, the students, and their parents, helped by apps like ClassDojo. Grade lev-els, in the manner created by Horace Mann, will go away; students will instead be grouped by factors like learning speed, level of indepen-dence, and social capabilities. Teachers won't stand up in front of the classroom and teach all the students the same lesson from curriculum developed by state officials; instead, they will become coaches and col-laborators, helping to guide projects, set goals, handle challenging sit-uations, and add a personal touch to online lessons.

Khan Academy got started when Sal Khan, after graduating from MIT, was working as a hedge fund analyst in Boston. In a story that's become something of a legend, he had cousins in New Orleans whom he wanted to help with schoolwork. In 2006 he started creating vid-eos, usually about math or science, by making drawings or writing equations and narrating over them. (The first one, about basic addition, was seven minutes and forty-two seconds long, narrated in a way reminiscent of *Sesame Street* shorts.) He posted the videos on YouTube so his cousins could readily access them. And he had an interesting realization: his cousins liked being able to rewind a video if they didn't get a point or to skip ahead if they were bored. They could learn at their own pace without the barrier or embarrassment of needing to ask someone to go over something again.

Because the videos were on YouTube, other people discovered them. By 2009 more than fifty thousand people a month were looking at Sal's videos. A year later it was a million a month, and by 2011 2 million visitors a month visited the Khan Academy video education site. By then Sal had found his calling. He wanted to build Khan Academy into a force that would reinvent education worldwide. Neither Sal nor I called it "unscaling," but in effect that's what he saw: a way to unscale education by making it less of a mass-production learning factory and more of a personalized experience, built to help each student learn what she wants at her own pace.

Much the way unscaling works in other industries, at first Sal built Khan Academy by renting scale on the first platforms of this new era of technology: he put out YouTube videos that could be accessed through the cloud on mobile devices so students could see them anytime and anywhere, even while inside a classroom. But now Sal's vision for the future requires the ur-force of the next wave of technology: artificial intelligence.

AI can learn about the individual student working on an interactive online course. The video course is evolving, with video instruction interspersed with exercises and questions. Because AI in the age of Amazon Alexa and Google Now can now understand natural language, students can respond verbally, much as they might to a teacher's question in a classroom. The AI can detect whether the student understands the material and then either go over it again or move on, always challenging the student enough to keep advancing the learning but not so much that the student gets frustrated and gives up. Over time the AI will understand what the student knows and doesn't know, her learning style and speed, and which subjects she loves and which she hates, and from this, it can then build coursework specifically for her. This is how education moves from standardized, one-size-fits-all curricula to education for a market of one in order to maximize everyone's potential. It is the difference between making all our kids conform to a mass-market education system and having an education system that conforms to your individual kid.

You might imagine, then, how Sal's vision of the K-12 classroom plays out with AI. (This is not just Sal's vision: many other organizations, such as Coursera, have been developing similar online courses.) The coursework becomes individualized, conforming to each student's learning style and pace. The other parts of schooling—social, community, collaboration, learning how to learn—are the work of the classroom community. They go hand in hand and result in people prepared not for last century's Industrial Age but for this century's Digital Age.

In Mountain View, California, I helped Sal build a primary school, called Khan Lab School, that would implement these ideas. Instead of grouping students by age and grade, they are grouped by level of independence and knowledge in certain subjects—so a student who is great at math but average in writing might be with older math students but younger writing students. Instead of students getting all the curriculum content from the teacher and textbooks, teachers instruct them in thinking and character skills—and some content—and students learn the rest of the content online or through their own research. Instead of being divided into subjects (e.g., math, history, etc.), the day is divided into self-paced work time and then group collaborative and hands-on project time. Students don't get a report card every quarter; instead, they and their parents get constant online updates and feedback. If you think about it, this kind of school is a lot more like the work environment at a tech startup, whereas the traditional school is more like work in a twentieth-century corporate office or factory.

Schools well beyond Silicon Valley are buying into these ideas too. Teachers in the Lawrence Public Schools in Lawrence, Kansas, ask students to watch Khan Academy videos for homework, and then in class students choose whether to work in groups on problems that put the Khan lessons into practice or to get help from the teacher. When SRI International studied such blended-learning (i.e., online and offline)

classrooms, it found that 71 percent of students reported enjoying using Khan Academy, and 32 percent said they liked math more since
they started working on Khan Academy courses. In fact, SRI found
that students who used Khan tended to perform better in math than
their standardized test scores would have predicted. According to the
study, "Positive relationships were found between Khan Academy use
and better-than-expected achievement outcomes, including level of
math anxiety and confidence in one's ability to do math." When the
teachers were surveyed, 85 percent reported that the use of Khan
courses had positively affected their students' learning and understanding of the material, and 86 percent said they would recommend
blended learning to other teachers. The report also said,

> Across the two years of the study, the majority (91 percent) of
> teachers indicated that using Khan Academy increased their abil-
> ity to provide students with opportunities to practice new con-
> cepts and skills they had recently learned in class. Eight in ten
> teachers also reported that Khan Academy increased their ability
> to monitor students' knowledge and ability, thus helping to iden-
> tify students who were struggling. Among teacher survey respon-
> dents, 82 percent reported that Khan Academy helped them
> identify students who were ahead of the rest of the class, 82 per-
> cent said it helped them expose advanced students to concepts
> beyond their grade level, and 65 percent, including 72 percent of
> teachers in schools serving low-income communities, said that
> Khan Academy increased their ability to help struggling students
> catch up.

Put it all together, and we're seeing that online, AI-driven coursework
like that from Khan Academy and community-building apps such as
ClassDojo are helping to dismantle the old standardized model and to
reassemble the idea of "school" as built for each individual student.

AI-driven coursework and unscaled school will also make us reimagine standardized testing. The very idea of standardized tests loses
relevance in an unscaled, market-of-one world. Instead, AI software
will comprehend what each student knows by how far she has advanced

in online coursework and the quality of the student's projects—in other words, the kind of feedback students and their parents are already getting at Khan Lab School. Eventually there may be no need to test in the traditional sense and no need to take the SAT or ACT to get into college—just permit the AI software to send the college admissions department a summary of your knowledge, capabilities, and learning style.

* * *

To a prospective employer a college diploma makes up for a lack of data about what's inside your head. There is no way for an HR official to tap a few keys on a computer and find out what you know and how smart you are. The diploma, then, is a filter. The college certifies, in a very broad way, that you know enough and are smart or talented enough to graduate from that particular institution. Employers in the twentieth century relied on that filter because they had nothing better. Yet as whole generations increasingly learn by interacting with AI, the AI will provide a better filter than a diploma. The AI will know precisely what you know and how smart you are, and any job applicant could give some version of that information to a prospective employer.

So why would you need a college diploma to get a top-tier job? If you are smart enough to tear through MIT or Stanford online engineering courses, the AI would be able to certify that you're as qualified as someone who spent four years and $200,000 attending those same schools. A diploma is a generalized statement about a broad group of people: everyone who graduated from this college is probably at a certain level of intelligence and capability—or were at least wily enough to get passing grades. An AI-rendered certification based on your online coursework would be a diploma for one—a detailed certification of *exactly* how intelligent and capable you are.

Almost everything about college is ripe for unscaling. In the United States scaled-up universities can't keep scaling fast enough to keep up with the growing global population seeking American schools. The cost of operating a scaled-up campus is soaring, while vastly more demand than supply allows for sky-high tuition. The cost to students has become a burden on society—and on businesses that need a growing supply of smart and knowledgeable new people to join the workforce.

Many top universities have been creating massive open online courses, known as MOOCs, to claim their place in an unscaled education market. By early 2017 four universities—MIT, Georgia Tech, the University of Illinois, and Arizona State University—were offering degrees based on MOOCs. One of the hallmarks of a MOOC is that millions of students can take, say, physics from the best physics professors in the world—and those professors are likely at one of the top-tier private universities. That threatens to disrupt many midlevel colleges. Why pay a lot of money to get taught by midlevel professors when you can go online for a fraction of the cost (or maybe free) and get that lesson from the best of the best? Not only that, but online AI will guide you by tailoring the course to your learning speed and style.

Any way you look at it, unscaled, AI-driven, college-level, personalized learning is about to challenge scaled college for the general population. To go further, the very idea of "higher education" is getting peeled out of the four-year, on-campus college experience. To get good jobs, people won't need to set aside a four-year chunk of their lives and a ton of money; instead, they might choose to learn a little at a time over a lifetime—perhaps even get many micro-degrees. Such a model fits much better with our twenty-first-century, digital-era work and careers. As we're already seeing, industries and technologies and demand for certain kinds of skills shifts quickly and constantly. To prosper people will have multiple, overlapping careers throughout their lives. Everyone will need to constantly learn in efficient and inexpensive ways.

Even online education proponents don't believe traditional colleges are going to disappear. But in the next twenty years the college landscape will shift greatly. A few great US brands—such as MIT, Harvard, and Stanford—will continue to be centers of research and places where bright young people can meet, mingle, and collaborate. Universities below that top level will have a tougher time in an unscaled world. Students in their market will increasingly choose some combination of online education and work to get ahead—and employers will increasingly welcome that.

* * *

Opportunities for entrepreneurialism in education are a bit different from for-profit industries such as energy and healthcare. For instance, although I'm involved in Khan Academy, it is a nonprofit, and I believe education as an institution should be nonprofit. However, I see a lot of interesting opportunities in technology—like that pursued by ClassDojo—driving the unscaling of education.

Here are some of those opportunities:

EDUCATION CLOUD: A decade ago everyone thought the big opportunity in education was going to be just putting videos of traditional college lectures online. But that has proven to be like putting radio shows on TV in the 1950s—video lectures on their own don't really take advantage of the technology. What's emerged since—from Khan Academy, Coursera, and others—is a combination of learning software, content, and human interaction in the cloud. Instead of watching a lecture on a screen, these courses operate more like a personal tutor, taking students at their own pace through multimedia learning. Instead of trying to emulate a traditional college degree program, many of the most interesting opportunities will be in lifelong education and nanodegree programs that help people learn very specific topics in a short time to advance their careers.

New companies will create education that flows to us instead of making us go out of our way to get it. The education cloud is meant to make learning easily accessible and tailored to our needs. Frankly, it is a huge market waiting to be filled. There are opportunities, for instance, to offer a company learning programs for their midcareer employees. We already see companies such as AT&T and L'Oreal signing up for online retraining with Udacity, an online learning company cofounded by former Stanford robotics professor Sebastian Thrun. L'Oreal employees can work through video-based courses to get a digital marketing nanodegree to further their careers. Still other companies have been developing platforms that let experts in anything from plumbing to singing offer their own courses online—just another kind of learning in the cloud.

The United States has sprinted out in front in cloud-based education, but it will soon be in demand in every major country as well. As of April 2017 only about 2 million people in India were registered for online courses—in a nation of 1.3 billion. Cloud-based education is nearly nonexistent across much of Asia and Africa. And, finally, I see a huge opening to create cloud-based education geared more to liberal arts. Much online learning today is aimed at math and science or at specific skills like software coding. In the AI age the very human thinking involved in writing, philosophy, history, and similar subject areas will prove to be more valuable than ever.

CONNECTED CLASSROOM TOOLS: As schools unscale, more teachers will adopt technology to create virtual classrooms that connect students, parents, and teachers to outside resources, educational content, experts, or anything else that can help students collaborate and learn at their own pace. ClassDojo and Edmodo are examples of all-around virtual classroom apps that help teachers manage these connected classrooms. Today it is a waste to have every teacher create their own lesson plans, so an online marketplace called Teachers Pay Teachers lets educators share their work with one another—an open marketplace where teachers can sell their lesson plans and other educational resources to other teachers. The more teachers can be entrepreneurial, the more they are likely to adopt new technology to help them do their jobs.

I can envision a whole new wave of apps that allow "schools" to be virtually built by connecting disparate classrooms all over the world. After all, a fifth-grade class of advanced science students in Kansas City would have more in common with a classroom full of similarly inclined students in Poland, India, and Chile than with the first-grade students down the hall in the same building. Add together mobile, social, cloud, virtual reality, and 3D printing, and classrooms thousands of miles apart could act as one. If that turns out to be a way to build new schools for a new era, it certainly makes more sense than constructing tens of thousands of new buildings from scratch.

VIRTUAL AND AUGMENTED REALITY: The possibilities for VR and AR in education are almost limitless. Tech giants such as Microsoft,

Google, and Facebook see this and, in response, have been developing experimental products. Microsoft's Holotours, for instance, are an early version of an immersive historical experience, allowing students to go back in time and walk around, say, ancient Rome. Google Expeditions has offered up virtual journeys to Mars or Antarctica. Of course, for more than fifty years students have been able to watch video of far-away places, but VR in the unscaled era enables learning to encourage exploration at the student's own pace. When VR allows students to go to ancient Rome, they can dive deeply into the culture of chariot races or the machinations of government. VR can also take students inside the human body to learn biology or inside the sun to learn about nuclear fission. These are only some examples of how VR can change learning, and a new generation of innovators will surely surprise us with concepts we can't yet imagine.

In the foreseeable future I see a different role for augmented reality. AR is a way for a student to stay anchored in the real world while some kind of lens—such as glasses, goggles, or a see-through phone—layers information or images on top. A history AR app might allow you to walk around a city and "see" it as it was a century before or to walk through the woods and see information about every kind of tree and plant. Years from now AR will effectively put two people who are thousands of miles apart in the same room, as if they were sitting next to each other. Imagine how this could transform one-on-one teaching: a student in the United States could learn Chinese from someone in China, conversing as if they were having tea together. If you think back to Sal Khan's first video tutoring lesson for his cousin—the video that turned into Khan Academy—it was a crude version of what AR will be able to do for students in the future.

6

Finance

Digital Money and Financial Health for All

I n the middle of the late-1990s dot-com boom, Ethan Bloch, then thirteen, started day trading stocks with his bar mitzvah stash. He did great, tripling his money in a couple of years. Of course, it was almost impossible not to do great in that raging-bull stock market. But when you're thirteen years old, it's easy to think you're king of the world.

In 2001, as the markets crashed, Bloch, by then all of fifteen, lost it all. "First, it taught me that I didn't know what I was doing," he says now. "And then it left a burning curiosity I still carry today, which is to understand finance, how it works, why we have it, what benefit we get from it, and what pain we get from it as well." He went to the University of Florida to study finance and psychology, determined to someday help people be smarter about their money than he was about his.

After college Bloch started a marketing software company that was acquired by DemandForce, a software company later acquired by Intuit, the pioneer of personal finance software. He could see that finance was increasingly digitized. If your bank accounts and bills are digital, you can feed them through software that—unlike the human mind—can make consistent rational decisions. And there certainly was a problem waiting to be solved, Bloch concluded. In 2013 the US Federal Reserve found that 60 percent of Americans aged eighteen to forty saved nothing. Adults under thirty-five had a savings rate of negative 2 percent—they were spending down their assets. "I had this basic idea, specifically grounded around the realities in America, which is if you want the majority of this country to have financial health, you need to make financial health effortless," Bloch says. To try to do that, he founded Digit, a cloud-based service that shows how banking is getting unscaled. Digit is another company that is part of the unscaling movement that I have made the centerpiece of my investment strategy.

Digit is software that handles your money in a smarter way than many people would on their own. It started with helping people build up savings. Once a user signed up with Digit she'd give it access to her bank account. The software would watch the money coming in and going out and then learn how much the user could save without feeling much pain. Then it would, a little at a time, transfer money out of her checking account and into her savings account—without asking, which turns out to be key: people save more when they don't think about saving.

Over time Digit has been getting more sophisticated. It can use machine learning to watch your cash flows and learn about your earning and spending patterns so it can help with more complex decisions. You can ask Digit, for instance, to help with a goal, like, "I want to save $2,000 for a vacation." It could plan accordingly and move more free cash to savings before you have a chance to blow it on expensive shoes. It can watch your bills and help keep in mind your savings goals and free-cash needs. Eventually, we expect, Digit will be able to be your financial agent. Just outline your financial goals and needs and then let the software learn your patterns and be your helper.

Why is Digit—and many other similar services—unscaling banking? Like everything else, banking scaled up in the twentieth century. It's how we got "too-big-to-fail" banks during the 2008 financial crisis. Big banks, though, focus on big customers and the mass market, standardizing their offerings to achieve economies of scale. They don't have much motivation to focus on creating innovative offerings aimed at twenty-somethings who struggle to make rent. Companies like Digit are starting to peel off those niche customers and serve them profitably, building products by renting scale on mobile networks, cloud computing, and AI. The more Digit's software gets to know an individual user, the more it can automatically tailor its service to that one individual—serving a market of one.

Digit is relying on existing banks as a platform so it can rent scale and capabilities from banks without needing to build a banking infrastructure and meet banking regulations. Users have an account at Wells Fargo or Bank of America, and Digit's software layers on top of that, in between the user and the bank account. It automates decisions and handles transactions. If you think about it, that's what the advisers and managers in a small-town bank branch used to do—get to know customers and help them make financial decisions. Companies like Digit are snapping off the bank branch and creating a new kind of consumer-focused business. "Every person's financial life is extremely calculable," Bloch says. "Every decision you make within that trajectory is extremely calculable. We all should have that service—we should all have that to lean on."

Bloch believes Digit will be a net benefit to individuals and even to society as a whole. "What happens when your finances are self-driving?" he notes. For one thing, people with Digit avoid overdraft and late fees, each adding up to around $20 billion a year in revenue for banks—most of it paid by those who can least afford it. "Think about the mental productivity lost by trying to figure out which bill to pay and how much, and how to make this other payment, and realizing you missed a payment," Bloch says. "All of that can be gone because everything's getting paid at the right time for the right amount. What happens to all that mental productivity we'll be given back? We'd already be on Mars if we had that!"

* * *

In the early twentieth century scaling up banks was difficult. Money was physical—paper currency, coins, gold bars. It couldn't be transferred to another town by wire but had to be stored in a safe, counted, and handed out through a teller window. Ordinary people didn't have credit ratings or a financial paper trail, so bankers preferred to lend money to customers they could know or easily learn about, probably within a short distance of the bank's office. Most businesses were small and local—the international corporation was still in its infancy. Farms were also family owned and small—not corporate-owned megafarms. Banking matched its environment. Most banks remained local, personal, and small. Although a few large banks, led by J. P. Morgan, were emerging, finance before World War II was generally a handcrafted business.

In the 1950s and '60s computers changed that by turning money into information. Recalling its history, the Federal Reserve Bank of Atlanta noted that in the late 1950s most banks were still running on mechanical tabulating machines, absent the electronic information of computers. "A visit to the check-processing department of a high-volume office like Atlanta or Jacksonville would mean walking into a room in which 70 to 85 women sat busily clicking away at gray, 36-pocket IBM 803 proof machines, punching in payment amounts and bank identification numbers with one hand and, with the other hand, picking up checks one at a time from a stack and pushing them into a slot," the bank says in *A History of the Federal Reserve Bank of Atlanta 1914–1989*. "A skillful operator could handle 1,200–1,500 checks per hour."

In 1963 the bank installed an IBM 1420 computer, which could process more than forty times the output of a human operator. Such developments led to economies of scale—a bank could achieve bigger margins by conducting more business with fewer employees, and a bank that could afford a computer or two could offer better deals to customers, undercutting smaller handcrafted banks.

Around the same time, corporations gained momentum and came to dominate the US economy. Banks scaled up to match their needs for capital and financial transactions. In 1950 Diner's Club offered the first universal credit card, then in 1958 American Express and Bank of America introduced credit cards as well. These new inventions enabled mass-market lending to consumers by using information about money and credit transmitted by telecommunications wires as a substitute for more intimate knowledge of each applicant. All this technology allowed a bank to serve standardized products to greater numbers of people and businesses. Economies of scale kicked in, and bigger banks became a better business model.

Yet one impediment kept bank scaling in check: regulation. Through the 1970s much of the federal regulation of the financial industry was still based on the pre–World War II business model—before computers, corporate hegemony, or credit cards. That changed in the 1980s, when legislation lifted a series of limits on banks and other kinds of financial institutions. The impact was dramatic. According to a Federal Deposit Insurance Corporation (FDIC) study, at the end of 1984 the United States had 15,084 banking and thrift organizations; by the end of 2003 the number had collapsed to 7,842. Almost all the disappearing banks were in what the FDIC called "the community bank sector with less than $1 billion in assets in 2002 dollars." Those were either acquired by scaling-up banks or were driven out of business by scaled banks. By 2014 a study by SNL Financial found that 44 percent of the total assets held by all US banks were controlled by the five biggest banks: JPMorgan Chase, Bank of America, Wells Fargo, Citigroup, and US Bancorp. In 1990 the five biggest banks held less than 10 percent of all US assets.

Thanks to the combination of technology, globalization, and the rise of the corporation, banks scaled massively, economies of scale won, and a few monster-size banks took over the industry. The bigger the banks got, the more they focused on high-profit, high-volume customers like corporations or super-wealthy clients and the more they needed to offer standardized, one-size-fits-all products to mass-market consumers.

In many ways scaled-up banking was good. It infused capital into business to help the economy grow; gave credit to millions of people so they could buy TVs, clothes, and other goods to improve their way of life; and helped millions get mortgages to buy homes. However, this scaling up also helped create the conditions for the 2008 financial crisis. Big banks had the incentive to offer loans to as many consumers as possible through products that could be standardized and securitized, which led to unhinged and irresponsible mortgage lending. When these loans started to fail, the biggest banks were "too big to fail"—the federal government needed to bail them out or they would've tanked the whole economy.

The 2008 financial crisis has been the topic of countless reports, studies, books, and even a Hollywood movie, *The Big Short*. Its causes are complex and often esoteric. But looking at it through my lens on technology and economics, I've come to believe the reasons for the crisis can be simplified: economies of scale had run their course in the industry. Decades from now the crisis will be seen as the beginning of the end of the dominance of super-scaled financial companies.

In the mid-2010s we're seeing the earliest glimmers of a new era of unscaled banking.

* * *

How will consumer banking unscale?

Companies such as Digit give us a glimpse of a new kind of consumer banking service. Digit is not a bank at all—it's a service built on top of banks, much the way an app on your phone is a thin layer on top of the massive computer power of AWS or other cloud-computing platforms. In that sense, giant banks like Wells Fargo are starting to open up their capabilities as a platform, becoming something of a banking cloud. Big banks can take care of the heavy banking duties, like maintaining highly secure financial computer systems and handling compliance with state and federal regulations—again, not unlike how AWS handles the tasks of building data centers and maintaining back-end software.

But the big banks will increasingly give up—or, more likely, lose—direct relationships with consumers. Instead, companies like Digit will offer products focused on the special needs of small segments of the banking market and will use AI to learn about each customer so the service can be personalized, getting closer to creating a financial service for a market of one.

We're in the early days of this shift. As of this writing Digit has only a tiny sliver of the banking market, but it's growing rapidly. If Digit is like a personal savings bank for young people, other niche banking apps will pop up aimed at—pick your target—parents saving for their kids' future, older people planning retirement, or immigrants who just arrived in a new country. I expect to see an explosion of innovation—products and services I can't possibly foresee. Someone will come up with a new AI-driven app that helps people pay for a house and skip the traditional mortgage, or a new way for AI software to help a newly married couple set up a joint account and budget their spending so they don't get into fights about money.

Banks have always set up a standard set of products and services—checking, savings, credit cards, loans—and expected us to conform to what worked best for *them*. That was standard operating procedure in the scaled era. In the unscaled era banking will conform to you. Banking apps will study you (with your permission, of course), figure out how you need to deal with your money in the context of your life and goals, and create a banking service just for you. It might not be anything like the banking service used by anyone else you know.

Consumers who need banking won't go looking for a bank. You'll instead go looking for an app that fits your unique financial needs. The service behind that app will have a relationship with a bank that will hold your money and comply with regulations, but you might not even know or care what bank that is. The AI behind the app will learn about your income and spending patterns, take into consideration your goals, and act as your personal financial assistant, doing all your financial tasks automatically—even constantly negotiating on your behalf for better rates on loans or lower fees for services. All you'd need to do is tell it if something changes in your financial needs. You could picture your banking app linked to Amazon's Alexa, so you might just tell Alexa, "I need to buy a new car. What can I afford, and how can I pay for it?" The app will give you an answer, get your instructions, and manage everything else behind the scenes. Even the credit score will go out the window—the app would know enough about you to already know whether you're likely to pay back a loan, so why bother with a lagging indicator like a credit score, which only knows how well you've paid off loans and bills in the past?

Consumer banking is simpler than commercial banking, but commercial banking is still going to change in significant ways as well. To see where that's heading, look at Stripe, which is peeling off business that had been badly served by megabanks' one-size-fits-all approach.

* * *

In 2010, while teaching a class at MIT, I first met Patrick and John Collison, charismatic students from Ireland. Patrick was an MIT student; his younger brother John was attending Harvard. They were starting a web-based payment processing company they then called /dev/payments. I asked Patrick who their customers would be, and he said the bulk of their customers haven't been born yet. The Collisons had a big vision for building a new platform for global online commerce. Their dream was that every new startup would use this service to get up and running quickly, serving markets anywhere in the world. That company became Stripe, which as of this writing is worth about $9 billion. As one of Stripe's original investors, I saw it as an AI-driven company that could unscale a chunk of commercial banking.

Stripe solves a real problem in this super-connected era. Before Stripe, setting up a business that could accept payments from other countries in other currencies was a bureaucratic tangle—one that PayPal or other apps couldn't fix. Banks would ask the company to fill out applications, often on paper. The process would take time. You'd need to fill out a new application to transact in every currency. And yet if your company didn't do this, it would lose customers outside the company's home country—if someone from the United States visits a website in Germany that prices in euros, most likely that person is going to just go somewhere else. Stripe automates the global payments process. (PayPal does some of what Stripe does, but PayPal is more of a third-party payments processor, whereas Stripe is more of a system that developers can build into any ecommerce site so transactions are seamless.) By the start of 2017 a company that signed up with Stripe could immediately start doing business in 138 currencies.

As John Collison—Patrick's brother and Stripe cofounder—puts it, Stripe is working to make money move as easily across the internet as any kind of data packet. One key to being able to do this is risk management and fraud detection—traditionally a function of banks. Stripe can offer it because of artificial intelligence. The AI learns patterns of transactions so it can spot possible trouble. Like all AI, Stripe's gets better as more data flows through it. As Stripe signs up new companies, it sees more and more transactions and traffic, feeding all that information to the AI software. So AI is helping Stripe profitably target a specific market—startups and small businesses—that get less attention from mass-market banks, which would rather do business with large companies.

Small companies could process payments before Stripe came along, of course, but big banks were more interested in big companies' business and didn't make payment services easy or cost-effective for small companies. But Stripe is a lean tech company built on top of banking platforms (i.e., using big banks to store and protect money) so it can profitably serve small-business markets that are more of an afterthought for big banks. Instead of making small companies conform to standard payment services, Stripe allows companies to configure payment services to their needs. In fact, Stripe is evolving into a company that can do much more than any bank. For instance, it can operate as a clearinghouse for marketplace platforms that need to be able to accept payments from buyers and distribute the money to sellers. (eBay is one of the best-known marketplace platforms.) Stripe has also introduced a service called Atlas, which helps startups anywhere in the world instantly set themselves up as a US-registered corporation with a US bank account. Because it's not a bank, it can put together various services to become a one-stop shop for starting a global unscaled company. Its killer application around payments earned Stripe the right to become *the* platform for unscaling finance. And now even big corporations are using Stripe—a sign that older scaled businesses will adopt newer unscaled solutions.

Fundbox, another company I've invested in, is peeling off a different slice of commercial banking. The company helps small businesses manage short-term cash flow gaps by advancing payments for outstanding invoices. Sometimes a small company might be waiting for a big payment but meanwhile needs to pay bills or salaries, so it needs a loan to bridge the gap. Fundbox uses AI to watch a company's accounting software and understand the risk levels of giving that company an advance. Then, when the company needs money against an invoice, Fundbox can make the decision in less than a minute and deposit the funds in the company's bank account the next day—an easier, faster, and cheaper solution than going directly to a bank.

Companies like Stripe and Fundbox are early signs of a revolution in commercial banking. Finance is now completely digital. Money is information on a network. Accounting is software. Transactions are automated. AI can see it all and put the pieces together to understand a company's financial picture and help it make decisions. One innovative startup after another is going to come into this space to serve narrow business markets profitably in ways big banks never could. With tailored services that use existing banks as a platform, these innovators will peel off commercial customers.

This is not necessarily terrible for the top-tier banks. Stripe generates business for banks that otherwise wouldn't happen, creating new business the banks wouldn't have. Everyone wins—there's more money flowing through the system. By enabling small companies to do more business globally, Stripe stimulates commerce, and by building its service on top of banks, it pumps new money through these banks. The difference, though, is that Stripe has the relationship with businesses and banks have a relationship with Stripe. Over time big banks will tend to lose contact with many end customers but still have a good business as the back end for many AI-driven unscaled services. Big banks will turn into big banking clouds.

In such a scenario banks would continue the consolidation trend the FDIC documented. Much like cloud computing services, the United States is only going to need a handful of banking clouds. On those clouds will be built an enormous number of banking apps and services that people and businesses rely on. Small and midsize banks—which don't have the scope to turn themselves into giant banking clouds yet don't have the limberness to compete for customers against banking app startups—will have less reason to exist. No doubt many will get acquired by the largest banks or simply fade away.

As with consumers, by the mid-2020s small and medium-size companies might not open accounts with a traditional bank; instead, they'll find services like Stripe and Fundbox that are aimed at their specific needs. Companies will sign up with those apps, and money will flow through the apps and into accounts with whatever banking cloud the apps use.

Bit by bit, what has long been scaled-up banking—where each bank is a bundle of mass-market services for consumers and businesses—will atomize. Services will get unbundled from banks and offered as stand-alone apps aimed at niche markets. We'll no longer conform to banks; banking will conform to us. It will be the biggest shift in the financial industry since the Atlanta Fed first installed IBM computers in the 1960s.

* * *

As you know by now, I work in a corner of the financial industry called venture capital. It has scaled up in its own way, especially since the 1980s, when the software industry took off. By the 2000s a handful of big Silicon Valley VC firms that could raise the biggest funds dominated the industry. Those firms were usually the first stop for an ambitious entrepreneur, so the top VCs saw the best opportunities first, which increased their advantage over small investment firms and ensured that the big firms would continue to use their version of economies of scale to build a massive competitive advantage.

In the 2000s unscaled investment models for startups started to emerge. First there were so-called angel investors, and then entrepreneurs Naval Ravikant (who had started Epinions) and Babak Nivi (who ran an investment blog, Venture Hacks) started AngelList in 2010. At first it introduced startups to angel investors and evolved into a kind of syndicate of investors, giving individuals some of the clout of big VCs. Around the same time Perry Chen, Yancey Strickler, and Charles Adler launched Kickstarter out of Brooklyn, New York. It became a way for almost anyone to invest in almost anything—a company, an arts project, a product—and was termed *crowdfunding*. By this writing Kickstarter has raised nearly $3 billion from 12.4 million people for more than 119,000 projects. AngelList and Kickstarter were serving niche investors and innovators in a way big VC firms never could.

In 2016 the US Securities and Exchange Commission instituted new rules dictated by the JOBS Act (Jumpstart Our Business Startups

Act). Policy plays an important role in unscaling, and the JOBS Act is an example. In the past, if you invested in a company on Kickstarter, in return you could receive only early access to the company's products or maybe a logoed coffee mug. If a startup wanted to offer equity, investors had to be accredited—they had to have an annual income of at least $200,000 or a net worth of at least $1 million. But the new rules instituted under the JOBS Act allow private companies to get financed through crowdfunding and, in exchange, give equity to the investors.

The old rules held back unscaling of venture investing. Now new kinds of crowdfunding sites are popping up to take advantage of the rules—sites such as SeedInvest, FlashFunders, and Wefunder. The home page of Wefunder gets right to the point: "Break the monopoly of the rich," it announces. "The wealthy enjoyed a government-protected monopoly on investing in high-growth startups. Everyone now has the right to invest in what they care about and believe in." Listings on the site range from a fiber-optic company to Rodeo Donut (tagline: "Gourmet donuts served with fried chicken and whiskey").

These democratized, unscaled investing mechanisms give more individuals a way to grab a share of the economy as well as make it easier for tiny niche companies to raise money and get started.

Now blockchain technology is threatening to take apart the financial industry's role in helping growing companies raise money. For the past century a company would raise money from private investors like me and later go public on a stock exchange to raise capital to fund growth. An initial public offering, or IPO, raises money for the company by selling stock in the company to investors. Big banks like Goldman Sachs dominate the IPO process. In the United States, IPOs are highly regulated, and going public has become costly and burdensome for companies.

In the 2010s blockchain technology opened up a new way for companies to raise money: the initial coin offering, or ICO. In an ICO a company offers "tokens" to the public instead of stock. These tokens are stored on a blockchain, so a token can be embedded with software instructions that dictate what it is. A token doesn't have to be a share of the company; it might include a promise to deliver a service or product, which is similar to the way fundraising campaigns work on

Kickstarter. ICOs create a lot of intriguing possibilities for companies and investors. Because the blockchain is essentially programmed to govern itself and tracks every transaction or movement of tokens, no central stock exchange is necessary. Because blockchains are distributed on computers around the world, an ICO is global. No nation's government could easily regulate it. A company could program its own rules into its ICO—rules that would be transparent to everyone. In the United States today all public companies must report financial results quarterly. An ICO company might instead decide it will reveal financial information weekly—or annually, or never—depending on how management wants to run the company. Investors get to see those rules and decide whether they like the idea of investing in such a company.

I believe ICOs are at a very early stage and may be a decade away from entering mainstream finance. In the first half of 2017, dozens of companies raised about $1.5 billion through ICOs, and celebrities such as boxer Floyd Mayweather and heiress Paris Hilton bought into ICOs. At the same time, governments are getting anxious about ICOs, which remain largely unregulated. The US Securities and Exchange Commission in mid-2017 issued a warning saying ICOs may run afoul of US securities law, and South Korea and China outright banned the offerings. Still, while it might take years, blockchain-based ICOs will likely undermine investment banks, stock exchanges, government regulators—and venture capital firms.

Venture-style investing—taking big risks on startups or innovations—will increasingly atomize. There will be more kinds of vehicles that give more kinds of people a way to back more kinds of entrepreneurs. As with banks, these are often people and companies that big VC firms could never serve all that well. But these unscaled investment vehicles can serve niche markets in a way that works for everyone. Big VC firms like General Catalyst, Sequoia Capital, and Andreessen Horowitz won't go away anytime soon, but our role will change. We're already becoming more suited to dealing with "big" startups that need a lot of capital, while entrepreneurs on the edges—those who don't need much to get started, are in parts of the country or world that don't attract much startup investment, or have a seemingly crazy idea—will more likely turn to the crowd or to angels for funding.

One thing is certain: as finance unscales, it will generate huge opportunities for new kinds of companies and business models. We'll need the right policies to help unscaled finance emerge in a beneficial way—and I'll get to that in Part 3. If all goes well, unscaling should end up spreading money and opportunity more evenly across the population, easing the divide between rich and poor.

* * *

Finance is a sprawling global industry that's ripe for reinvention. In many ways it hasn't caught up to the revolution of the past decade brought on by mobile networks, social media, and cloud computing. It's still an industry of monolithic institutions running on mainframe computers and private networks. But the forces of unscaling will work on finance and banking as surely as they are changing other industries.

Here are some opportunities I see for entrepreneurs:

CONSUMER APPS: Banks have not really offered consumers innovative new ways to deal with their money. Now companies are rushing in to insert themselves between big banks and consumers, essentially using the banks as platforms and winning the relationship with consumers. Digit is one example, and I see many others emerging as well. Wealthfront and Betterment, for example, offer consumers AI-based financial advice. The user allows access to his or her financial accounts, and the AI software can learn patterns and goals and then offer the kind of direction that used to come from human financial advisers. The AI financial advice costs a lot less than paying a professional. Stash is an app that helps consumers invest in stocks a little at a time, depending on their personality type. A socially conscious investor can be guided to put money behind good causes. Lenddo relies on data from social networks—instead of a traditional credit score—to determine whether someone will repay a loan, which means the company can lend money to consumers who have never had a bank account or credit card; most of its customers so far have been in developing markets such as the Philippines and India. As banking unscales and AI allows software to learn about individuals, more services will arise that can offer new kinds of banking to the unbanked—which, the World Bank says, is more than 40 percent of the global population.

One other interesting consumer opportunity concerns taxes. In another decade it will seem silly for anyone to do their own taxes or hire an accountant to do them. For many people, almost all their finances are digital—bank records, IRA accounts, credit card transactions, and so on. And the tax code is a giant algorithm that can be encoded in software like TurboTax. So now enter AI, which can understand your finances as well as the tax code and then automatically figure your taxes constantly, instead of once a year. By the time tax day comes along you'd already know exactly where you stand. Just click a "send" button, and you'd be done. In early 2017 IBM and H&R Block started offering Watson AI-based tax help for Block customers. At this point the AI helps guide Block's tax pros, who then work with consumers much as they always have. But it's not hard to see where this is going. Before long, some version of AI tax technology will be put right in consumers' hands through the cloud, and tax preparers and traditional tax software will have trouble keeping up.

COMMERCIAL SERVICES: Big banks offer an array of services to big companies, but small businesses can have trouble getting their attention, much less a loan. Unscaling means that entrepreneurs can profitably focus on niche markets, so there's no question we'll see a flood of new companies offering interesting, focused commercial services for small business.

Fundbox, mentioned earlier, is one example—a company that uses AI and digital financial data to make instant decisions on advancing small companies money against their invoices. Stripe is perhaps the most powerful instance of this, creating a payments system that can help a small company anywhere in the world instantly sell globally. Lendico, based in Germany, bills itself as a "marketplace for loans" for small businesses. Completely cutting out banks, the app connects borrowers to anyone who has money to lend, using AI to analyze the reliability of both the borrower and lender. Think of any kind of financial service that a small company needs, and it will wind up embedded in an app in the cloud, powered by AI.

DIGITAL CURRENCIES: Bitcoin, Etherium, and other blockchain-based global currencies got a lot of attention in the 2010s. Although I'm not convinced we're going to disrupt national currencies with these crypto-currencies anytime soon, policymakers need to pay attention. China watchers believe that nation is working to develop a crypto-currency that could undermine the US dollar's dominant position in the world economy. If crypto-currencies gain widespread acceptance, they might displace the national currencies of smaller nations or nations with unstable currencies. These would be unprecedented developments, and we need to understand their potential ramifications.

Beyond currencies, blockchain technology allows for money, trust, and contracts to be encoded in software and distributed across the planet to anyone with access to the cloud—in ways that are tamper-proof. In other words, the money can't be stolen, trust can't be falsified, and contracts can monitor themselves and signal all parties if some promise has been broken. Here's one example of an innovative company rethinking the idea of "money" by using blockchain: Plastic Bank has a mission to give people—primarily in developing nations, where large chunks of the population don't have bank accounts and carrying cash can make you vulnerable—an incentive to collect plastic for recycling. The company partnered with recycling collection centers around the world and with companies that make products with recycled plastic, tying the ecosystem together with software to help create a global market for recycled plastic. On the other end, anyone can sign up with Plastic Bank, collect plastic, sell it into Plastic Bank, and get blockchain-driven digital currency delivered to their cloud account, which can be accessed via mobile phone. The digital currency can be used to pay for education, healthcare, or other needed items. There's even a way a user can go to certain bank ATMs and turn their Plastic Bank currency into cash without having a bank account. The combination of all these technologies—blockchain, mobile, cloud—will drive thousands of innovations as imaginative as Plastic Bank.

7

Media

Content You Love Will Find You

The media industry has unscaled more than most—making it an industry that others can learn from. Our unscaled media have brought us many new-era wonders, such as Netflix movies on demand and Spotify streaming music. But it has also damaged journalism—an institution critical to the US system—and ushered in media echo chambers that have contributed to divisive politics all over the world. Our current media are an example of what can happen when policymakers fail to think through the implications of unscaling.

In this century mobile, social, and cloud technologies have done quite a job disaggregating and unbundling media. In journalism individual articles live on their own, separated from their home publications like the *New York Times* or *Newsweek*, and get reaggregated on Facebook or Twitter, which then get ad dollars that had once gone to the publications. Network TV clips of Stephen Colbert antics land on YouTube, separate from his late-night TV show. Streaming services like Netflix and Hulu, which let subscribers choose to watch any show or movie at any time, had more US subscribers by 2017 than cable TV, which has traditionally bundled together a vast array of entertainment. In music, songs on streaming music services such as Spotify and Tidal float free from albums. Along the way the internet era has let loose millions of blogs, video shows, do-it-yourself songs, and podcasts.

And yet AI and the economies of unscale have only begun to play out in media. Oddly enough, radio—long one of the less-sexy media sectors—provides a surprising window into how the dynamics of the AI century might further change journalism and entertainment, taking it from mass media to audience-of-one media. The hope is that coming developments will help bring about a profitable and productive era for media.

Let's start with how radio scaled in the past. Getting into the business had been expensive. It cost a lot to secure radio spectrum for the broadcast, build a radio tower, buy broadcasting equipment, hire DJs and news personnel, and build a sales team to sell the ads that bring in revenue. Once a radio station owner had all that up and running, she would be spending nearly the same amount to reach ten people within reach of the station's signal as a million. So the way to make money in radio was to create programming that appealed to a broad audience, get as many listeners as possible, and raise advertising rates. The next-level step was even better: accumulate radio stations, serve standardized music playlists across all of them so fewer personnel would be needed, and centralize ad sales and other business departments. Then, toward the end of the twentieth century, US policy changes helped accelerate scaling in radio. The Telecommunications Act of 1996 ended key rules that prevented any one company from accumulating radio stations and taking undue advantage of economies of scale. Before the 1996 Telecom Act no company could own more than forty radio stations. By the 2010s the biggest radio company, iHeartMedia (formerly Clear Channel), owned more than twelve hundred. The lure of economies of scale set in motion a wave of consolidation as soon as it was allowed, leading to a decline in the diversity of radio programming as companies like iHeartMedia sought to standardize and automate music playlists.

But shortly before Congress enacted the 1996 act, radio began experimenting with the internet. In 1994, for the first time, a radio station—WXYC FM in Chapel Hill, North Carolina—fed its broadcast into the internet, allowing anyone anywhere in the world to hear it on a computer. A radio "station" no longer needed a radio tower to exist, only a URL. Dozens of startups took a shot at internet radio. In 1998 Mark Cuban, now the colorful owner of the Dallas Mavericks NBA team and star of the TV show *Shark Tank*, and his lesser-known co-founder Todd Wagner were swept up by the idea that radio broadcasts of college sports could be pumped through the internet and reach fans who might live thousands of miles away. They formed a company, eventually called Broadcast.com, to aggregate internet radio (and later, video, though that didn't work very well during the dial-up era of the internet). Broadcast.com excited Yahoo so much that it paid $5.7 billion in 1999 to buy it.

Yet under Yahoo Broadcast.com fizzled out and disappeared. Internet radio was having little impact and making no money. In those days of low-fidelity cell phone networks, an internet radio station couldn't yet reach cars or handheld devices. Advertisers had no idea who was listening. They were long used to at least knowing listeners were in a defined geographic area, so they could try to sell them on local restaurants or car dealerships. But advertisers had no idea what to do with an audience scattered globally. On top of all of internet radio's challenges, consumers had a hard time finding what they wanted. Internet radio programming was too hard to search.

So the internet created a promise of unscaling, which people like Cuban and Yahoo's management spotted early, but they couldn't deliver on the profits and business models to make unscaled radio viable. For that, the emerging unscaled radio industry needed a new kind of rentable platform.

This is where an entrepreneur in Texas named Bill Moore came in. He'd gotten an MBA from the University of California, Berkeley, and worked for a half-dozen years at Efficient Networks, a company that sold software to telecommunications firms. Around 2002 he had an idea to build a TiVo-style digital recorder for radio, which could be programmed to record and save any radio show. That idea immediately highlighted the search problem for radio—there was no good way to search online for radio content. Google could crawl the words in text-based news stories to serve up search results. A DVR attached to cable TV could look up the TV listings to know what show to record. But radio had few listings because it is made up of spoken words and music, which search engines couldn't crawl. If you were looking for a radio show that would play a certain kind of music for a couple of hours on a certain day, good luck—there was no way to find it. Radio was stuck in a previous era, a medium seemingly unsuited to the internet.

To address some of these challenges Moore decided to create what he called Radio Mill. He hired contractors all around the world to enter information about local radio stations and their programming, and Moore put the information on software that consumers could install on a PC. A user could type in, say, NPR's *This American Life*, and the software would find the next station that was going to air the show, record it, and save it to the hard drive. This was, in effect, Moore's TiVo for radio.

As a consumer offering, Radio Mill bombed. But Moore kept morphing his concept. He opened up Radio Mill's API so other developers could use it to build services or build it into consumer audio devices. The iPhone hit the market in 2007 and mobile networks ratcheted up to broadband speeds, making high-quality internet radio portable. Now independent radio could rent scale to compete with scaled-up giants. By 2010 Moore moved his company to Silicon Valley, where it was reborn as TuneIn.

Over the next few years TuneIn turned into something of a Netflix for radio. It had searchable data on thousands of stations worldwide, and its app could let a user find and listen to any radio content. The company assembled rights to major league sports audio broadcasts. By 2016 it had grown to 60 million users. Pretty much any user anywhere—in a car, at work, at home, on a walk—could use TuneIn's data to find and listen to any radio content at any time at a level of quality similar to traditional radio broadcasts. A focused radio outlet could rely on TuneIn's platform to find and serve a niche audience of like-minded listeners all over the world, peeling off parts of the audience otherwise owned by giants like iHeartMedia. And just as Netflix uses data and AI to inform decisions about programming it produces, TuneIn creates some of its own content based on what the AI says about its listeners.

All this led to AI and the real unscaling revolution emerging now. The new technology meant internet stations could gain an advantage:

terrestrial stations know relatively little about each individual listener. TuneIn, however, is getting a flood of data about what users listen to through the app, where they are when they listen to it, and what kinds of people tune in to what kinds of stations. That data, sifted through TuneIn's AI-driven software, can recommend stations to users much the way Amazon can say that if you like this knife, you'll probably like this kind of spoon, or the way Netflix recommends movies based on what you've previously watched. In this way TuneIn can help a small independent station reach the right passionate audience and know a lot about who is in that audience. All of this has helped internet radio make money for the first time. Over the next few years that should create a positive spiral for internet radio, says John Donham, now TuneIn's CEO. "The quality of content on digital radio will skyrocket and eclipse that of terrestrial radio," says Donham. "When [radio] really makes this transition to being an online medium, it's hard for me to imagine a future in which the terrestrial tower is the anchor for live audio." Radio trade publication *Radio Ink*'s advice to the terrestrial industry echoes Donham's view. "The most successful audio programmers will find ways to marry the best of what analog radio offers with the portability and personalization of the digital environment," wrote Stacey Lynn Schulman, head of strategy for Katz Media Group. A giant radio corporation serving lowest-common-denominator programming to the most people possible will increasingly have trouble competing against focused programming that seems to speak directly to each individual as an audience of one.

What's happening with radio is similar to how AI and economies of unscale are taking apart and reassembling every form of media—after more than a hundred years of media scaling up.

* * *

In the early years of the twentieth century "media" mostly meant newspapers. The number of newspapers published in the United States peaked in 1909 at twenty-six hundred. Almost all of them were independently owned—a collection of small companies and family businesses, each rooted in one city or town. In New York more than fifteen daily newspapers were being published in 1900, ranging from the *New*

York Times to the *New York American, Daily Mirror,* the *World,* the *Sun,* and the *New York Journal.* (Today it's down to the *Times, Daily News,* and *Post,* unless you count the national *Wall Street Journal,* which is based in New York. Most American cities now have just one newspaper.)

The then-new technology of efficient, electricity-driven automated printing presses set economies of scale in motion at newspapers. One of the chief movers was Hermann Ridder. Born in New York to German Catholic parents, he developed a breakthrough printing press while running a German-language newspaper as well as a Catholic newspaper. In 1911 he founded the International Typesetting Machine Company to make his press called the Intertype. The first was installed at the *New York Journal of Commerce* in 1912, and Intertype became one of the most significant suppliers of newspaper printing presses in the decades to come.

Such presses made it possible for a single company to print many more newspapers in a shorter time and reach a larger daily readership. A bigger customer base would bring in more money from advertising and subscriptions, improving the economics of operating a press and employing reporters and editors. The path to more subscribers was news content with mass appeal, so newsrooms sought to cover broader topics of wider interest, and they started bundling together every kind of news a town might find useful or interesting—sports, business, politics, crime, comic strips, and so on. Newspapers that could scale up had a competitive advantage over newspapers that could not. By the 1920s vast numbers of newspapers that couldn't scale were going out of business or merging, and successful newspapers grew larger. From 1919 to 1942 the number of US newspapers fell by 14 percent, according to *Columbia Journalism Review,* even though the US population jumped by 29 percent. Surviving newspapers saw their circulation surge. The age of scale in media was in full swing.

The same dynamic took hold in every kind of medium in the first half of the twentieth century. The first federally licensed commercial radio station—KDKA in Pittsburgh, Pennsylvania—went on the air in 1920. In 1926 the National Broadcasting Company (NBC) turned on its national radio network, followed by Columbia Broadcasting System

(CBS) a year later. The networks' concept was that the bigger the audience, the better the economics. Getting that bigger audience meant shooting for mass appeal, whether covering news or baseball games or broadcasting popular dramas such as *The Lone Ranger*.

Television came next. The first experimental TV station, W2XB, broadcast from a General Electric factory in Schenectady, New York in 1928. Eleven years later, in 1939, NBC began its first regular TV service with a telecast of President Roosevelt's speech at the New York World's Fair. US consumers quickly embraced television in massive numbers. In 1941 about seven thousand TV sets were in existence in the country, by 1950 the number had grown to nearly 10 million, and by 1959 an astonishing 67 million TV sets were in American homes and businesses. Once again mass markets and economies of scale ruled. TV programming aimed for the widest audience by producing safe, broad comedy like *I Love Lucy* and by showing popular sports such as baseball and boxing. As TV news developed, it sought to be nonpartisan and objective so it would appeal to everyone.

By the second half of the twentieth century TV, radio, and newspapers reached almost every American household. By then the path to greater economies of scale was consolidation—owning more newspapers or TV or radio stations meant a company could more efficiently reach an ever-growing mass market by taking advantage of economies of scale. Regulatory rules designed to prevent a few companies from owning too many outlets loosened as the decades wore on, all but disappearing by 2000. The newspaper company Hermann Ridder had founded merged in 1974 with a newspaper company founded by the Knight family to become Knight-Ridder, briefly the largest newspaper publisher in the United States. Gannett, started in 1906 by Frank Gannett in Elmira, New York, went on a buying spree and, by 1979, owned seventy-eight newspapers across the country (and it owns more than a hundred today). iHeartMedia bought up hundreds of radio stations. Television similarly consolidated. In many cases the same companies bought up newspapers, radio, and television outlets to form giant media conglomerates. In 1983 about fifty companies shared in most of the US media ownership; by 2012 ongoing consolidation had shrunk the number of super-dominant media entities to a handful that included Comcast, Disney, Viacom, News Corp., Time Warner, and CBS. The 2010s seemed to be the apex of scaled-up media giants.

Yet the 2010s also marked the moment when AI and the economies of unscale started giving big-media executives migraines.

* * *

The technologies developed since 2007—mobile, social, cloud— whipped open the door for unscaling media.

Newspapers bundle coverage of many different topics and a variety of advertising—one package for a broad audience. Sites like Craigslist and Monster.com peeled off classified ads. Blogs and focused news sites like the political site Politico (founded in 2007) and business news site Business Insider (2009) started peeling away niche audiences interested in more stories about those topics. Little by little, new, more focused online outlets pulled away chunks of the newspaper bundle. And then social media—Facebook in particular—yanked individual stories out of print news outlets; instead, people could post stories for their friends to see. Facebook's algorithm could learn what kinds of stories you were likely to read and then essentially serve you a reaggregated mix of news from a variety of sources, all suited just to you—a market of one (for both better and worse, which I will get to later).

These early years of unscaling turned into an economic nightmare for legacy print news companies. Advertising drained out, and stock prices plummeted. New York Times Co. stock in 2017 was worth less than half its 2002 peak. In 2014 print and online media companies employed twenty thousand fewer journalists than twenty years before, according to Pew Research. In 2016 alone Pew counted four hundred buyouts or layoffs at newspapers—a trend that shows no sign of abating. My coauthor, Kevin Maney, lived through this transition. He wrote for *USA Today* for twenty-two years and left in 2007, just before the newspaper's first of many rounds of buyouts and layoffs.

Television no longer commands the mass-market audience it used to. Not only are viewers scattered among niche cable channels, but technology has allowed companies like Netflix and Amazon.com to compete against broadcast TV with on-demand programs. On You-Tube, where anyone can make a show in their bedroom and offer it globally, some of the top stars can outdraw traditional network TV sitcoms. (Lily Singh's Superwoman channel, which consists of funny bits like "Why Bras Are Horrible," has 10.3 million subscribers and earns $7.5 million annually.) Video content—whether full-length movies, live sports, news, or self-produced vignettes—is now disconnected from the TV set and is just as likely to be watched on a smartphone, tablet, or laptop. Pew Research found that the percentage of smartphone owners who have used their phone to watch movies or TV through a paid subscription service like Netflix or Hulu more than doubled from 15 percent in 2012 to 33 percent in 2015.

I mentioned earlier that I first met Snapchat's founders (the parent company is called just Snap now) in 2012. Snapchat got its start by letting users send photos that quickly disappear. It then built its business model on the idea that it is unnatural to generate data about everything we do. For thousands of years before the internet, a conversation disappeared the second it was over, no device logged everywhere you went, and when you finished the newspaper, the newspaper didn't know which stories you read. Snapchat struck me as private communication the way it was meant to be. That was the moment of my "big aha," when I understood, as mentioned earlier, that technology will finally conform to us rather than the other way around. This is now becoming important in media. For too long our habits bent to the ways that media was distributed. It was never particularly natural to schedule yourself to be in front of a TV at the precise time your favorite show was to come on or to read the news only when the paper arrived in the morning and sift through all the stuff you didn't care about to get to the stories that mattered to you. Even in the internet age many media outlets like Netflix or Facebook soak up data about you so they can serve up more of what you might want—but that's not natural either. For thousands of years before the internet, you didn't need to give up your privacy to be entertained.

So if Snap started as a natural way to have conversations, it can also become a model for a more natural way to serve up media—whenever you want it, without any tracking of your media habits, redefining media for mobile devices. It is becoming its own niche—and media companies such as CNN, the *Economist,* and Vice are paying Snap to be on it. Snap makes its money not by selling ads but by renting its platform to media companies that want to reach a young, mobile audience that doesn't want to be tracked. As I write this, Snap has about 160 million active users around the globe. Although it is losing money, the Snap platform is still in its infancy, and I believe it has a good chance of becoming an important business.

Step by step, platform by platform, scale as a competitive advantage in media has been getting neutered by post-2007 technologies. The industry has been atomizing. Small, focused upstarts can build on global platforms like YouTube, TuneIn, or Snap and serve niche audiences that once could only turn to mass-market offerings from media giants. Yet still, technologies of the next decade are about to create entirely new kinds of media opportunities.

Artificial intelligence will play an increasingly huge role in media. AI promises to help consumers find exactly the news or entertainment they want and guide each piece of news or entertainment to the right audience. Think of it as Netflix on steroids. Netflix circa 2017 recommends movies or shows based on what you've watched and how you rated it. As mentioned, Netflix also gathers data about its audience's viewing habits to help the company know what kinds of content to produce. (Netflix is certainly showing the way to do this well. A share in 2012 was about $12, and in mid-2017 it was around $150. Amazon and other streaming services are now copying Netflix's strategy.) As more content gets consumed online, every media choice becomes a data point. Software will be able to know who's viewing what, where they're viewing it, and even what else they're doing while viewing. On the extreme—and controversial—end of that trend, a company called TVision Insights can use a TV set's camera to monitor the person watching the TV and send back data to advertisers and programmers about whether the viewer's eyes flicked down to her phone during commercials or whether she was smiling or frowning during a show—yet more data to better make highly targeted shows and ads. Viewers need to opt in for TVision to be able to grab the data, but the existence of the technology raises concerns that some other entity might invade your media privacy without you knowing.

In an AI-driven media world the media we want will literally find us, even if it's some obscure program produced on another continent. The AI will know us so well that it will only serve us media we will really like, and programming will be waiting for us on whatever device we want, perhaps geared to the time of day and where we happen to be (because the GPS on our phones will know). Media will truly be aggregated for an audience of one. Bundled networks, channels, or newspapers won't make sense. We'll increasingly want the AI to assemble a "channel" that is just for each of us, culled from every kind of media produced anywhere by anyone. Again, though, this raises the concern that we'll all wind up in our own media silos instead of sharing media experiences with the mass market. That's good or bad, depending on your point of view. Media consumers will presumably be happier because they'll get what they like. But it also threatens to isolate each of us in our own media worlds.

AI holds the key to media profits. The most valuable advertising online today is the most targeted. Advertisers will pay more to a Facebook or Google because it can learn about you from your activity and fire ads at you that you'll likely want to see. AI-driven media platforms will take that to the next level. If you opt in and let your media access data from your online and offline activity (as captured through IoT devices), you'll only see ads for products you're likely to desire—and only see the kinds of ads that are effective on you. If you respond better to funny ads, that's what you'll get. The more personal the ad can be to you, the more an advertiser will pay for it. This way a startup can make a TV show in a garage that will appeal to a niche audience, serve it up on an AI-driven media platform, reach exactly the right audience, and get paid by advertisers seeking the perfect target—and content gets more diversified and profitable at the same time. This is where TuneIn is heading—a rentable AI-driven platform to help internet radio find the right niche of listeners so advertisers will pay to reach their target customers. It's a sign of the times for media.

In such a media market scaled media loses its advantage. People want entertainment and news that's just for them, and this will often win out over mass-market programming designed to lure the broadest audience. There will always be some content that's appealing *because* much of society is tuning in, such as shared experiences like the Super Bowl or Oscars. But the days are long gone when millions watch a bland sit-com because it's the only thing on. Most of the time we will each be huddled in our own media bubble, consuming content created by every type and size of media outlet. Big media companies will lose more and more of their audience to upstarts, and traditional outlets like network TV will continue to lose to platforms like Snap, Netflix, and YouTube.

The idea of AI-curated personal media bubbles brings up a highly relevant issue. The 2016 election of Donald Trump as president revealed a severely divided nation, and the 2010s media landscape contributed greatly to those divisions. When economies of scale ruled in the news media, it made economic sense for the major media to produce centrist news for the broadest market. When economies of unscale rule and news media gets fractured, however, the better business model is to target a small, passionate audience, play to personal biases (see Fox News and Breitbart News Network), and peel viewers away from the centrist mass-market news outlets. That dynamic is behind the rise of highly partisan outlets such as Breitbart and the loss of faith in traditional mainstream media. A 2016 Gallup poll showed that only 32 percent of the public trusted the mainstream media. In 2017 President Trump continued to label mainstream media "fake news" to try to discredit it, at least among his ardent followers. As I write this in mid-2017 it's difficult to tell whether this dynamic will continue or snap back in a counter-trend.

There's little doubt that unscaling will keep driving people to narrower, more biased media as well as media entrepreneurs to create news outlets aimed at profitably serving passionate niche audiences. We will all feel like we're getting news that's made just for us, but each of us will be getting news that's different from everyone else's news. These bubbles will make it harder for us to bridge divides and talk to one another. If past is prologue, this will continue to fracture our politics and roil our scaled-up major political parties. Whether that's good or bad depends on your point of view.

* * *

The media industry began to unscale earlier than most, and the first wave opened up opportunities for new kinds of focused news operations (Politico, Buzzfeed), new data-driven entertainment companies (Netflix, Hulu), and many new ways for us to get music (from iTunes to Spotify). The next set of opportunities is likely to be built around artificial intelligence—and, in coming years, around virtual and augmented reality.

AI PLATFORMS: TuneIn evolved into an AI-driven platform that can enable profitable unscaling of a medium. As described above, AI can learn about individual consumers' tastes on one end and content creators on the other and then provide a way to deliver content—and the advertising that pays for it—to people who passionately want it, no matter where they are on the planet. TuneIn does that for radio. Netflix, in its own way, does it for TV series, acting as a platform for show producers to reach the right audience.

I believe we will continue to see opportunities for AI-powered platforms for news, music, books, games, or any other kind of content. Amazon's Echo—a speaker-like gadget that responds to voice commands using the company's Alexa AI—is a new kind of platform that's driving experimentation in delivering audio news, music, audio books, and other information. (Google, Samsung, Apple, and other companies are rushing to keep up with Amazon.) Ford and other automakers are imagining the self-driving car as a new on-the-road media platform—a place where people might watch movies or location-based media served up through AI. We're only at the beginning of AI-driven media platforms, and no doubt some will emerge that will take us all by surprise.

VIRTUAL REALITY: In virtual reality certainly one direction will be immersive entertainment. As the technology gets better, someone will figure out how to make a movie that we're a part of instead of watching it on a screen. Someone will figure out how to render live sports in VR so we're experiencing it on the field, as if we're a player. Someone will figure out how to allow us to stand on a virtual stage with U2 or Beyoncé and play along with the band.

Those are self-enclosed experiences. VR will really get interesting when it becomes an open platform, more like the internet, where you can wander and exist much like in the real world. Philip Rosedale, who founded one of the first virtual worlds, Second Life, is working to make his new company, High Fidelity, into that kind of networked VR platform. In such a world shopping at an online store would be like a cross between walking the aisles of a physical store and the instant shopping of an Amazon.com or Wayfair. Imagine all the things you can do on today's internet—work, socialize, learn, shop—rendered instead in VR so it feels almost like doing those things in real life.

AUGMENTED REALITY: In another decade AR might become even more interesting than VR. To truly create new kinds of entertainment, the technology will need to improve—and it's been challenging to create true AR that can seamlessly insert moving images into the real world around you. Ideas for AR abound among developers. We might see some new AR media company take public figures—former presidents, famous authors, top academics—and encode their personalities, mannerisms, and speech into AR so that anybody could have them virtually over for dinner. Perhaps a whole movie or videogame could be set in your house or neighborhood. It's easy to imagine a cooking show that blends with what you're actually doing in the kitchen. For that matter, any kind of instructor could be virtually with you, giving you guitar or ski lessons in a way that mimics a private lesson. It's a good example of an unscaled medium moving toward a market of one in a way that's automated and profitable.

We're seeing some cool early experiments. One developer set up a murder mystery game in a house in North Carolina. Players moved through the house using their smartphones like Sherlock Holmes's magnifying glass, looking for clues that were all AR overlays. Something like that could be a powerful blend of human interaction and virtual technology.

If you add all the developments in media together, the unscaling of media will, in turn, abet the unscaling of mass-market consumer products, which is what the next chapter details.

8

Consumer Products

Everything You Buy
Will Be Exactly What You Want

By 2010 mobile, social, and cloud platforms—and the ability to rent these platforms to instantly scale—had created a sense among entrepreneurs that new, unscaled companies could be created in *any* kind of business. That's when a group of graduate students with little funding or experience concluded they could challenge a powerful global consumer-product giant and upend the experience of getting a pair of eyeglasses. Their thinking led to Warby Parker.

Neil Blumenthal, one of Warby's cofounders, at least wasn't a complete rookie in the business of eyeglasses. As an undergraduate in international relations and history, he took a fellowship in El Salvador and, while there, worked with VisionSpring on its unusual philanthropic model. VisionSpring trains "vision entrepreneurs" in developing countries, who then travel the countryside checking eyesight and selling glasses for less than $4. Blumenthal helped these traveling entrepreneurs do a better job selling glasses and even spent time in China trying to source glasses at lower costs.

Blumenthal then enrolled in the Wharton School's MBA program, where he met David Gilboa. Just before starting at Wharton,

Gilboa—who'd been a pre-med undergrad student and gone on to work for a while at investment firms Bain Capital and Allen & Co.— left a $700 pair of Prada glasses in an airplane seat back pocket. Thinking about the $200 he paid for the iPhone in his pocket, Gilboa recalled, "It didn't make much sense that I was going to have to pay over three times what I paid for the phone for a new pair of glasses." At Wharton Gilboa related his story to Blumenthal and two other students, Andrew Hunt and Jeffrey Raider. (Raider would later found Harry's, the men's grooming startup with a Warby-like model.) "We started talking about why glasses were so expensive," Gilboa told *Fast Company*. "Then we learned a little bit about Luxottica."

Luxottica is exactly the kind of corporation that great business leaders built in the twentieth century. In 1961 in a town in Northern Italy, Leonardo Del Vecchio founded Luxottica to make spectacle parts. By 1967 he was selling complete glasses and, a few years later, began making glasses for other brands. He was a believer in vertical integration and the economies of scale. So he bought an eyewear distribution company to get his product to the marketplace. He cut licensing deals with designer labels. He bought retail stores. His strategy was to create the biggest company in the business, selling to the largest mass-market aggregation of customers possible. The key was to sell the most glasses so he'd take the greatest advantage of his fixed costs, making Luxottica more efficient and profitable than any smaller competitor and stacking the deck in Luxottica's favor.

By the 2000s Luxottica ruled the $65 billion-a-year industry and made glasses sold under famous brands like Armani, Oakley, Ray-Ban, Prada, Chanel, and DKNY. It operated stores such as LensCrafters, Pearle Vision, Sunglass Hut, and Target Optical. (In early 2017 Luxottica doubled down on the scaled approach by buying French lens maker Essilor for $49 billion, renaming the company EssilorLuxottica.) In eyewear Luxottica became the master of being all things to all people. That was how you built a superpower in an era when scale won. It took decades, and it created a competitive advantage because it would take a long time for a challenger to get big enough to be dangerous.

Beyond that, Luxottica was deeply invested in a certain kind of consumer experience of getting eyewear—built on older physical platforms such as roads and shopping malls. That experience went like this: first you had to get your eyes checked by an ophthalmologist or optometrist and get a prescription; then you would take that prescription to a store in a mall, try on frames, and order the frames and lenses; and finally you'd drive back to that store a couple of weeks later to pick up your new glasses.

While Luxottica continued to scale up and invest in the old experience from end to end, the technology industry was building powerful digital platforms that included the internet, mobile networks, smartphones, social media, digital payment systems, cloud computing, open-source software, and global contract manufacturers. Each of those platforms gave almost anyone access to capabilities that Luxottica once had to build by itself. For example, whereas Luxottica had to develop worldwide retail chains to sell glasses to a huge market, the internet plus mobile networks plus cloud computing plus digital payment systems meant an upstart like Warby Parker could set up an online store that changed the game overnight.

Which is what Warby did.

The core idea was to sell a curated selection of hip, designer-style glasses online for a fraction of the cost of Gilboa's lost Pradas. Customers could order five frames to try on and send back—a twist on Netflix's DVD-by-mail model. (Eventually Warby started opening physical stores so people could browse glasses.) Warby took advantage of new platforms. On the internet it could open a store, which could sell to anyone anywhere, for next to nothing. Facebook and Google gave it ways to find customers and for customers to find Warby for far less than buying ads on TV or in magazines. The company could host its business in the cloud, never needing to buy a server or rent a data center. It could hire contract manufacturers to make batches of Warby frames. And when commerce moves online, it generates data. Whereas Luxottica might need to do market research to find what customers want, Warby could get constant feedback from its online stores and aim products specifically at its urban, young-generation target market.

Luxottica could get data about what stores are selling, but Warby would get data about what individual customers are buying—or even, on the website, what they're looking at but not buying. Such intimate information about customers helps Warby know what designs appeal most to its target market. Once Warby later began opening retail stores, data from the website helped it know where to locate stores (so the stores would be near dense populations of Warby customers) and what to put on those store shelves.

The power of the unscale dynamic can be seen as Warby Parker developed its business model by *renting* scale instead of building and owning it and, in a flash, competed against Luxottica for a slice of the entrenched company's global market. The company can *rent* computing power on cloud services like AWS and Microsoft Azure, *rent* manufacturing from contract factories in Asia, *rent* access to consumers via the internet and social media, *rent* distribution from delivery companies like UPS and the US Postal Service. Warby today can succeed against an entrenched player with fewer than eight hundred employees. The company as I write this is worth well north of $1 billion and has become a fixture in the market for hip eyeglasses.

Warby is also part of a trend that is changing consumers' relationships with brands. Brands were created to convey information about products at a time when it was hard for consumers to get information. But our hypernetworked and data-engorged era is killing the very reason for mass-market branding. We can find out everything about some gadget or shirt or hockey stick from a maker we never heard of. We can read reviews, Google the company, ask about it on social networks. As we get better information about small-scale products, people feel safe seeking out the unique, the undiscovered, the unbrands—giving a company like Warby an opening against a Luxottica.

The result is a societal shift, as authors Itamar Simonson and Emanuel Rosen describe in their book, *Absolute Value: What Really Influences Customers in the Age of (Nearly) Perfect Information*. We used to want the brands everybody else had. But now we're moving toward mass individualism, wanting stuff nobody else has. This leaves big brands vulnerable to hordes of quirky little unbrands. Hilton becomes vulnerable to Airbnb's one-of-a-kind dwellings. Microbrews take share from

Budweiser. Tiffany becomes vulnerable to jewelry makers on Etsy. And Warby snapped cleanly into the middle of this trend. It wasn't just Gilboa who started wondering whether it's worth paying $700 for glasses just for the Prada brand cache—a whole generation who grew up with the internet has been thinking that way, and this is opening consumers' minds to unscaled products and services.

The economies of unscale are reshaping a great deal about the consumer experience. You can see how different it is to get glasses from Warby. You order online, try on frames at home, pick what you want, and your glasses get delivered to your home. You get what might've been $700 glasses for $100. Traditional advertising and marketing are nowhere in the picture. You still need to go get your eyes checked first, but even that part of the experience is changing. Startups such as Opternative are working on online eye tests that would give you a prescription for glasses. Fold that into the Warby experience, and the entire process of getting glasses would completely bypass everything Luxottica built. Luxottica won in its era by taking advantage of economies of scale. Warby is winning in the new era by taking advantage of the opposite—the economies of unscale.

* * *

The consumer market has been the most powerful economic force globally for the past hundred years. Consumers worldwide spend about $43 trillion buying goods and services—about 60 percent of global GDP. In the United States the $11.5 trillion consumer market makes up about 71 percent of GDP. Amazingly, last century's economies of scale created pretty much the whole of mass-market consumerism. Before that, a mass consumer market really didn't exist.

Technology platforms introduced in the early 1900s played an enormous role in making it possible to create and serve a mass market. The birth of national radio and television networks gave companies a way to reach nearly every consumer with advertising, blanketing each city and town with the same ads—which had to be as plain and broad as possible to appeal to all kinds of people. The arrival of cars and trucks and the expansion of the highway system meant that the same mass-market product could be distributed to stores everywhere, and consumers could drive to stores and malls to get the products that advertising convinced them they desired. Mass electrification coursed through all these developments, powering radio and TV, keeping lights on in stores, helping to automate assembly lines, and opening up new possibilities for breakthrough products, from vacuum cleaners to power drills to model trains.

By the second half of the twentieth century all these tech platforms—transportation, mass media, communication—supported the emergence of supermarkets, McDonald's, 7-11, and, ultimately, the epitome of mass-market consumerism, Walmart. Big nationwide or worldwide outlets needed to fill their inventories with products from big suppliers that could make a lot of what the masses would buy. Power shifted to companies like Procter & Gamble, Anheuser-Busch, Nike, Sony, Luxottica—companies that could buy expensive advertising, build huge factories, and churn out products that appealed to the wide middle of the market. A small-scale shop that catered to a niche in its own town didn't stand a chance against a Walmart stocked with name brands at low prices, even if the shop was a much better experience.

All the dynamics of mass-market consumerism reinforced one another. For instance, big producers could claim more shelf space at big outlets, making it difficult for niche products to reach consumers. And economies of scale ruled. The bigger the retailer, the harder it could bargain for lower prices and the more efficiently it could operate. Walmart especially gained an advantage in this way. The bigger the consumer product maker, the more it could spend on advertising while gaining efficiencies from mass production and massive distribution. Here, P&G took the lead ahead of other manufacturers.

But when you think about it, the mass-market consumer companies made each of us conform to the experience that was best for them, not us. Few of us want to drive miles to a Walmart, find parking, walk through Spartan aisles to try to find what we want (with the understanding that it might not be there), deal with staff who don't know us or our preferences, stand in line at a cash register, load our purchases into the car, and drive back home. On top of that, the products we wind up with probably don't conform to our specific taste. We're often buying a compromise product—Budweiser beer, Levi's jeans—designed to appeal to as many people as possible. The experience most of us really want is to easily find exactly the product we're looking for, no matter where we are at the moment, and have it delivered into our hands within a couple of hours. That's an unscaled consumer experience.

Amazon led the way toward creating this new consumer experience. Jeff Bezos founded Amazon on the idea of creating the "Earth's biggest bookstore." We were used to physical bookstores, and a big Barnes & Noble would carry about two hundred thousand titles, a huge number but still a fraction of the available books in the market. If you made the trip to a bookstore hoping to find something very specific, there's a good chance you'd be out of luck. Bezos reasoned that on the internet he could make every published book available in one global store and then ship it to any customer's address. We as consumers started to see how the experience should conform to us—sending the exact book we want right to our door—instead of our conforming to the sellers and producers by accepting what they were offering us.

Over time, of course, Amazon expanded beyond books into every kind of consumer product and also added AI that could tap the vast amounts of data flooding in from shoppers to help match the right product to the right person—honing in even further on a market of one. Robots find and ship products from Amazon warehouses, and its delivery network can get goods to consumers sometimes within hours. Today it's relatively easy for any quirky product to rely on the Amazon platform to reach a target audience that might be scattered across the planet. Walmart, spread among stores all over the world, sells about 4 million different products; Amazon sells more than 356 million. Amazon made us expect to be able to use a phone or computer to find anything and order it.

Seeing Amazon's success, competitors leapt in. Google has tried to become a platform for shopping and same-day delivery. Seamless, Deliveroo, Postmates, and Uber, with services like UberEats, want to become a last-mile delivery platform for small businesses and restaurants. At the same time, the mobile, cloud, and social platforms have made it easier to start a consumer products business and sell niche products to the world. Put all the platforms together, and it set the stage for Warby and other new-era consumer products companies such as Dollar Shave Club and The Honest Company to effectively compete against the scaled-up consumer goods giants. The economies-of-scale defenses that Luxottica or P&G built around their businesses—with manufacturing, access to consumers via shelf space, and mass-media advertising—had become vulnerable.

By the mid-2010s the consumer market was unscaling at an accelerating pace. Consumers increasingly wanted that unscaled, market-of-one experience. You could see the results in retail. In 2017 nearly a quarter of the malls in the United States—310 of the nation's 1,300 shopping malls—were at risk of losing their anchor store, according to commercial real-estate firm CoStar. In 2016 shoppers aged eighteen to thirty-four bought more apparel on Amazon than through any other retailer, accounting for nearly 17 percent of all online clothing sales to this demographic, more than double the market share of second-place Nordstrom. P&G found almost every one of its brands under assault by unscaled competitors such as Freshpet, Julep Beauty, and eSalon—all selling a focused line of products online to a niche market. Freshpet, for instance, sells only "healthy" pet food, and eSalon asks you to fill out a profile about your looks and coloring and then sends you custom-created hair coloring treatments.

And that's only what's developed so far, built mostly on the internet and mobile, social, and cloud technologies. The next wave of AI-driven technology will alter the consumer experience and the consumer industries even more radically.

* * *

As smart, new-era retailers and consumer-product makers interact with customers, they gather data that helps AI software learn about each customer, tailoring the consumer experience to each individual. Essentially AI automates the concept of personal service, making it economical to offer that kind of attention to everyone. And who doesn't want personal service instead of Walmart generic clerk service? It's a winning formula.

Stitch Fix is one glimpse into the coming consumer experience. The company, founded by Katrina Lake in 2011, is melding AI and human input to offer a personal shopper experience to anyone. As a new Stitch Fix customer, you start by filling out a style profile online, telling Stitch Fix your size and shape and also a bit about your lifestyle. ("What's your occupation?" "Are you a parent?") Then the company sends you five clothing items chosen by a stylist, based on your profile. You buy the ones you like and send the others back—an act that teaches the AI more about your style. Every few weeks (you set the interval) Stitch Fix sends a new batch of clothes, and again you respond by either buying or sending the items back. Through these transactions the software comes to know your style and, over time, can guide the stylists so they send you exactly what you like.

The Stitch Fix experience goes well beyond today's Amazon experience, which helps you find any product you want. Stitch Fix instead gets to know you and sends you products you might not yet know you'd want. The AI helps predict and anticipate consumer desires. That approach will spread throughout the consumer market—to groceries, makeup, home décor. Software will anticipate your needs and send products that specifically suit you. Such a consumer experience will be devastating to mass-market brands. We'll let the system send us products that best fit us, whether those products are made by a giant like P&G or a startup in a small town on the Saskatchewan prairie. The products you order might be delivered by drones or small autonomous delivery vehicles, getting stuff to you so quickly, negating the need to drive to a store for instant gratification.

Food, which became a highly scaled industry over the past century, is on the verge of going through a data- and AI-driven transformation. To feed more people, create national brands, and fill giant supermarkets, the family farm gave way to the corporate farm—huge tracts of land maintained by farm machines the size of houses. The trend kicked into high gear during the 1950s and 1960s, according to federal statistics compiled by Wessels Living History Farm. From 1950 to 1970 the number of farms declined by half—consolidated or sold to create bigger farms. Over those twenty years the average size of a US farm doubled. Yet the number of people on farms halved, dropping from more than 20 million in 1950 to fewer than 10 million in 1970. And productivity increased—farmers produced more food at a lower cost on roughly the same amount of farmland. Although those trends were most dramatic in those decades, they continued through the next forty years.

Now technology is starting to unscale the farm and move food production closer to consumers. One example is a farm, Bowery Farming, inside a New Jersey warehouse just fifteen minutes outside of New York City. LED lights mimic natural sunlight. The crops grow in nutrient-rich water beds on trays stacked floor to ceiling. And the farm estimates it can grow a hundred times more greens per square foot than an industrial farm. IoT sensors constantly monitor the plants and environment and send the data back to AI-driven software, which can learn what's best for the plants and tweak lighting, water, and fertilizer to improve yields. Much of the "farming" is done by robotic machines that move or water plants. Many kinds of urban farms—inside buildings, on rooftops, on random patches of land—are increasingly found in cities all over the world.

A startup called Lettuce Networks is using cloud and mobile technology to network urban farms. Founder Yogesh Sharma even calls it an Airbnb for farming. The company contracts with owners of small plots throughout a city and installs sensors that can monitor crops and environment. Nearby residents can subscribe to the Lettuce service to get food delivered. So the system knows what's being grown all around the city and, from that, assembles a basket of local food for delivery. Owners of the plots make some money off their harvest, while subscribers get an assortment of fresh food grown nearby—much fresher than food grown on farms thousands of miles away and shipped in refrigerated trucks. If food networks like this work, each city or town can become more self-sustaining.

Once again this all leads to a consumer experience that conforms to us rather than the other way around. For decades now we've been trained to get food at the supermarket—a task that involves driving somewhere, spending valuable time walking aisles, choosing mass-market products that are good enough but not exactly what we'd want (especially the tasteless industrial-grown tomatoes and berries), and loading the car for the drive home. In the unscaled era a system like Lettuce Networks could know your taste, know what's available locally, and deliver it to you fresh. For anyone old enough to remember a milkman coming to the door every few days and delivering just the right items, the future is looking more like the past than the present.

In fact, if you put all this together—the developments in consumer products, retail, food, manufacturing, brand value—it points to a consumer experience in the 2020s that looks very different from that of the past fifty years. Most of what we go out and shop for today will instead come directly to us. We'll sign up for different services—a Stitch Fix or Lettuce or whatever—that get to know us and deliver what we want when we want it. Common chore shopping for groceries or staples should almost disappear. We'll go to physical stores primarily for entertainment or education or to see new styles or products that push the boundaries of our usual purchases. We'll usually be able to find products that seem custom tailored to us and will never have to settle for a mass-market brand. We'll feel more pride in our quirky nonbrand purchases than in buying some designer label or high-end model.

This change will have an enormous impact on the industries that built the last century's consumer experience. We're entering an era when brands are devalued and advertising has less impact on consumer purchases, generating massive implications for legendary companies like P&G, Coca-Cola, Apple, Nike, Anheuser-Busch InBev, and Louis Vuitton. The decline of brands could be devastating for mass-market national media, especially television. Big-store retailers, whether Walmart or Safeway or Best Buy, will see traffic dwindle as more purchases move online and to AI-driven subscription services.

In short, the consumer experience built on scale over the past century is about to get disassembled and unscaled. The advantages of big are waning. In this new era small, focused companies that put the consumer at the center will beat big, mass-market operations most of the time.

All this will have vast ramifications for real estate, land use, and cities. The real estate firm Green Street predicts that at least 15 percent of shopping malls will close within a decade. Walmart alone in 2017 was operating 3,522 supercenters in the United States, each up to 260,000 square feet in size. As more consumers order products to be delivered, some of those Walmarts will close. Think of all the real estate that will be freed up and the opportunities to use that space. Already we've seen shuttered malls become condos, health clinics, hockey rinks, and giant indoor greenhouses. At the same time, patterns of movement of people and products will change as unscaling plays out. We will move less, so fewer cars will jam roads and fill up parking lots at predictable times, which will free up yet more land. Conversely, more goods will be sent to us, so city planners will need to apply new thinking to how to accommodate delivery trucks and, eventually, small self-driving delivery robots—not to mention delivery drones whizzing above the streets.

* * *

The great opportunities in the consumer market will revolve around giving every individual exactly what he wants, when he wants it. It reflects the constant theme in unscaling: scaled-up, mass-market products have long made us conform to *them*, but unscaled products and services conform to *us*. They will seem like they are built just for each one of us—customization built with automation. Over the next decade we'll see innovators transform one kind of product after another, moving them from mass markets to markets of one.

Here are some of the opportunities I see:

UNBUNDLING THE GIANTS: Consumer product companies from P&G to Nestlé to Samsung were built on the mass market. A hit product was one that appealed to the greatest number of people—one size fits most. But mass-market products are a compromise for most consumers. They're not exactly what we might want, but it's good enough and easily available. And that leaves an opening for small, new companies that can use technology to create products that hold great appeal for narrow slices of the consumer market—consumers who will feel like that product was created especially for them.

We've seen The Honest Company, Warby Parker, Dollar Shave Club, Blue Apron, and a growing number of other companies do just this. Honest was founded by actress Jessica Alba after she had her first child and had a hard time finding natural and organic baby products. Once she found business partners Brian Lee, Sean Kane, Christopher Gavigan, and Maria Ivette, the company started by making more environmentally friendly disposable diapers, competing against mass-market brands like Pampers. It now markets a range of natural household and beauty products. Warby focused in on a section of Luxottica's customer base. Dollar Shave Club, which sells blades by subscription, sliced off part of Gillette's market. Blue Apron, which delivers meals to cook at home, took a bite out of the prepared frozen food market. Just about any kind of mass-market product will be vulnerable to an unscaling strategy. We'll see this play out over and over again.

And these new companies will have the opportunity to rebundle groups of narrow-market products into the new P&Gs, as we're seeing with Honest. These new companies might get big, but they will always be more of a collection of businesses highly focused on serving a particular kind of customer.

OMNI-CHANNEL STORES: Through the history of civilization people have been drawn to markets. We like to shop. For many people it's a social and entertainment experience as much as a search for a product. So no matter how much commerce moves online, it's not likely that retail stores will disappear. But retail will certainly change. Successful retail stores will be part of a complete experience that connects online and offline shopping.

Warby Parker's stores are a way for customers to try on hundreds of frames in a hip atmosphere, yet most will buy the style they choose online later. Amazon shocked the book world when it started opening physical bookstores—the very kinds of stores Amazon helped kill over the past couple of decades. Why do this? As it turns out, people enjoy browsing bookshelves, and it gives them ideas about books they want to buy later online. As long as Amazon sees its stores as way to generate demand for Amazon shopping, it almost doesn't matter if an individual store is profitable.

Successful retailers will use AI to get to know customers and then tailor in-store experiences to the right set of customers in a city or town. A clothing shop might invite customers who have a specific sense of style to an exclusive session with a stylist who specializes in that look—then build an online shopping experience that plays to those customers' taste. The store and site work together to drive business and make that set of customers feel special. The more data they have on each customer, the more the in-store experience can be leveraged to make each one feel like a market of one.

LOCAL FARMING: Scaled-up farming has fed the world, but it's also given us "fresh" tomatoes that taste like plastic. A host of technologies, from AI-controlled grow lamps to IoT sensors that can constantly measure nutrients in soil, are making it feasible to profitably grow food indoors near customers—the farming equivalent of distributed manufacturing. In cities those indoor farms can be in old shopping malls, factories, warehouses, or parking decks. As everyone knows from farmers' markets, if food is grown nearby, it doesn't need to endure shipping—so it can ripen the way it's supposed to. In the middle of winter indoor-grown local tomatoes will taste like tomatoes. As you can imagine, that's what consumers will prefer.

Many new companies like Bowery are entering this space. Freight Farms is growing food in cargo containers. New York–based Brightfarms says it "finances, designs, builds and operates" greenhouses close to food retailers and has raised $11 million in funding. Edenworks is building rooftop greenhouses that grow greens, mushrooms, and herbs with the help of manure from tilapia and prawns also grown in the mini-farm. In Minneapolis Dave Roser, a former Hewlett-Packard executive, runs Garden Farms, converting multiple warehouses into indoor farms. Expect to see many more startups like these. Over the next decade local fresh food will eat into the market share of corporate farms.

ROBOT DELIVERY: This sector might take longer to develop, but a couple of decades from now the UPS guy is likely to be replaced by some kind of automated delivery vehicle. It's hard to say what kind of vehicle might work best. Amazon is famously working on drone delivery, and there's a good chance that smaller items or food—which is time sensitive—might be brought to you in a self-piloted drone. Once self-driving cars get good enough to operate without anyone in the driver's seat, some version of a robot UPS van might pull up to your driveway and text you to come out and get your package. Starship Technologies, an Estonian company run by two of the founders of Skype, has built a six-wheeled robot about the size of a baby carriage that can navigate sidewalks to deliver a pizza or cheesesteak from a restaurant down the street. Delivery service Postmates has been testing them in Washington, DC, and San Francisco. As I was writing this, Wisconsin state legislators were passing a robot delivery law that would limit the robots' operations to sidewalks and crosswalks and set an eight-pound weight limit and a speed limit of 10 miles per hour.

Although robot delivery might seem strange today, it will come to seem normal in this unscaling era and will be an important part of the grand shift from mass markets to markets of one. Instead of your going to Walmart to buy things everyone else has, a robot is going to pick up a product manufactured just for you in a nearby distributed factory and drop it at your feet.

PART 3

CHOICES FOR
A GOOD OUTCOME

9

Policy

Profound Decisions to Have a Good AI Century

Chris Hughes is hyperaware of the way technology has created inequality in America. He grew up in a working-class family in the small town of Hickory, North Carolina. His father was a paper salesman and his mother a schoolteacher. In 2002 Hughes started his freshman year at Harvard University and was assigned a dorm room with Mark Zuckerberg and Dustin Moskovitz. Those three, plus Eduardo Saverin, founded Facebook the following year. In 2016 *Forbes* put Hughes's net worth at $430 million.

Hughes and I are friends, and from my experience he has never forgotten that he won the business lottery. He left Facebook in 2007 to work on then US senator Barack Obama's presidential campaign and tried (though ultimately and spectacularly failed) to revitalize an institution of journalism, the *New Republic*. By the mid-2010s Hughes was focusing on what he believes is a coming crisis: automation driven by AI and robotics, coupled with other social dynamics, could soon leave vast numbers of people with no way to make enough money to lead a dignified life. He has been a driver of the Economic Security Project, which raised $10 million to fund two years of research into the idea of a universal basic income, or UBI. (I'm involved with the project and help to support it financially.) The project wants to figure out whether it's a good idea to give every citizen a regular payment so they can lead a decent life even if they are automated out of a job.

If we add up all the changes AI and unscaling will bring, what are we creating for society and business? Will this be heaven—or some version of hell? Hughes and many others in technology and government are thinking about how to avoid the latter—as am I.

The AI technologies and forces driving economies of unscale are inevitable. We're not going to stop artificial intelligence from happening—genomics, 3D printing, robotics, drones, IoT, and everything else I've described are already here. In turn, those technologies will continue to unleash dynamics that tear down the twentieth-century economy of scale and replace it with the new century's economy of unscale. This process has been in motion for a decade and will play out with gusto over the next ten to twenty years. The automated accounting and banking in the electronic commerce platforms like Stripe will leave behind millions of finance professionals and contract lawyers. New manufacturing based on 3D printing promises a massive shift of jobs away from big company factories to smaller, automated, on-demand factories. These trends can't be ignored. Policymakers need to grapple with how to help people make the transition. Our leaders need to look ahead and ask critical questions about the outcomes we desire from today's technologies. If, a century ago, we could have foreseen the impact of the internal combustion automobile—on the landscape, energy, air, cities, wars—would lawmakers have passed different kinds of regulations or incentives? No doubt!

Technology is neither good nor evil—it's what we do with it that dictates how it will impact society and the planet. The global technology industry needs to embrace responsible innovation. We've so long been focused on "disruption" that we need to be reminded that morally bankrupt disruption can leave a society in shambles. There are signs the tech sector is starting to get it. IBM, Microsoft, Google, Amazon, and Facebook came together to form an organization called Partnership on AI, vowing to show some sense of moral duty. "We recognize we have to take the field forward in a thoughtful and positive and implicitly ethical way," Mustafa Suleyman, the group's cochair and cofounder of Google DeepMind, said when the group was formed. Hopefully these companies will follow through and the rest of the industry will rise to the challenge. But it's critical that policymakers do the same. It can't wait. We're already ten years into the unscaling revolution, with the next two decades of enormous change bearing down on us. The time to make sure this revolution turns out well is now.

As I write this book two areas I'm interested in are autonomy (e.g., driverless cars and robotics) and longevity (i.e., how to extend our lifespans). Applying autonomy to truck driving seems ideal for creating a multibillion-dollar company. Longevity seems like another multibillion-dollar idea. Our firm, General Catalyst, invested in one company, Elysium Health, that makes a pill that can help repair DNA in ways that can help you live a longer, healthier life. I see quite a few companies working on both ideas. If you put the autonomy and longevity developments together, it raises a daunting question: What if we stick the landing on both? Many people will have less work while living longer. This is not just a potential problem for truck drivers and factory workers; many professional jobs may get automated away. In 2014 Goldman Sachs invested in and began installing an AI-driven trading platform called Kensho (General Catalyst is an investor). In 2000 its US cash equities trading desk in New York employed six hundred traders, but by 2016 that operation had two equity traders, with machines doing the rest. And this is before the full brunt of AI has come into play at Goldman. "In 10 years, Goldman Sachs will be significantly smaller by head count than it is today," Daniel Nadler, CEO of Kensho, told the *New York Times*.

Dramatic shifts in jobs, wealth, and power go hand in hand with technological revolutions. This happened in the early twentieth century as jobs, wealth, and power shifted to brand-new industries like cars and electrical appliances and to new professions such as engineering and professional managing. And the technology wave of today is shifting jobs, wealth, and power to new kinds of AI-driven companies in healthcare, autonomous transportation, and clean energy.

Such grand revolutions upset lives and careers and leave people behind. A sense of disorientation and frustration were evident during the

2016 presidential election of Donald Trump and the Brexit vote in the United Kingdom to exit the European Union.

Historically, new kinds of automation have always created more jobs—and better jobs—than they've destroyed. The twentieth century was a time of fantastic automation—it ushered in everything from the assembly line to the computer, demolishing countless obsolete jobs. At the same time, from 1960 to 2000 it created the job of computer specialist. In 1960 there were about twelve thousand of them; in 2000 there were 2.5 million, according to the Bureau of Labor Statistics. The number of accountants and auditors soared from thirty-nine thousand in 1910 to 1.8 million in 2000. The number of US college officials and faculty totaled twenty-six thousand in 1910 and grew forty-three times, to 1.1 million by 2000.

Will AI-driven automation similarly create jobs—or destroy them? We don't yet know. And the question of what to do if large parts of the population have limited opportunities to work is why Hughes and others such as Sam Altman of Y Combinator have launched their basic-income research projects. We need to figure out if it makes sense to give everyone a monthly check—and how to do it. Finland, the Netherlands, Switzerland, the United Kingdom, Canada, and Holland have all explored basic income as well, according to the Economic Security Project. One small-scale precedent is Alaska, which pays about $2,000 to every resident from the state's oil revenue.

But there are difficult questions about basic income to explore. How would it get funded? Estimates from believers in basic income programs say it would require an additional income tax ranging from 17 percent to 36 percent, and those against it say the additional tax would reach 60 percent. Would those tax rates kill the economy—or would the redistribution to millions of people who would spend the money speed up the economy? Would government's bureaucratic burden explode as they identify and pay every citizen a stipend—or decrease because the stipend would end the need for other programs like welfare? These are the questions we're asking in the Economic Security Project, and they are the questions lawmakers need to debate right now.

The basic income conversation doesn't account for what work has provided for centuries: structure and purpose. When I was growing up, the conversation at the dinner table often involved the question, "What do you want to do when you grow up?" Most people want to *do* something with their lives—they want to feel like they matter and contribute in some way, whether through work, art, family, or faith. More thinking needs to be done here. We should think about evolving technologies through the lens of unleashing human potential rather than replacing it—for instance, most AI scientists believe the technology can become our partner, like a brilliant assistant, helping people do better work. That should be the goal for AI, instead of using it solely to automate away jobs. And for the work that does get automated—conversations around basic income are a good start in addressing automation's impact on employment and livelihood. Still, overall we need to seriously start exploring what fulfillment and happiness look like if we create a post-work world for a sizeable chunk of the population.

* * *

In addition to concerns about jobs, there are two systemic issues with AI and unscaling that concern me. One involves monopoly platforms, and the other is algorithmic accountability.

MONOPOLY PLATFORMS

Monopolies might seem like an odd worry, as you've just read this whole book about how scale no longer equals power. But there's one wild card in the size equation, and it centers on how artificial intelligence software learns. As I've said, unscaling is possible because entrepreneurial companies can rent scale and profitably address global markets while remaining small and focused. But it's also true that those unscaled companies will need to rent their scale from platform companies, and some of those platforms will get huge. And when it comes to platforms, in most cases one winner vacuums up most of the market share, leaving competitors far behind. One example today is Amazon Web Services, the dominant cloud computing platform. It is key to unscaling, hosting the operations of thousands of entrepreneurial companies that challenge big corporations—and that has allowed it to achieve massive scale. Competing cloud platforms from Microsoft,

Google, IBM, and others struggle to keep up. Why? Because the more companies use AWS, the more data it gathers about how companies use cloud computing, which helps Amazon's AI learn how to better serve its customers, which in turn creates a bigger gap between AWS and its competitors that have less data coming in. This kind of dynamic is going to play out over and over in the AI era. Platforms that take the lead will have more data than competing platforms, thereby allowing those platforms' AI to learn much better than competitors' AI. This in turn gives the leading platforms an advantage, resulting in leading platforms winning yet more business, gathering yet more data, and gaining an even bigger lead on competitors—a spiral that can result in the leading platform becoming a near monopoly.

As platforms gain power, they tend to extend the platform, bundling in services that make the platform even more attractive to users while blocking out potential competitors. Google is a highly visible example. The Google search platform, in its mission to suck up as much data as possible, extended to things like Maps, Gmail, Docs, and so on. Users love having all those offerings integrated and working together, and the Google platform abets unscaling by giving entrepreneurs an easy way to rent—or get for free—a lot of functionality they used to have to buy or build. But the more Google (or its parent, Alphabet) can bundle together, the more the company reinforces its near monopoly in search.

We can end up with monopolies controlling key platforms that thousands or millions of small companies rely on—small unscaled companies that couldn't exist without being able to rent that scale. Then what if a monopoly started jacking up prices? Or using its position to wipe out challengers or to squash innovation—all things that monopolies sometimes do? Some of these platform monopolies are already among the most powerful companies in the world. As of this writing the five most valuable US-based companies are all in some way digital platforms driving unscale: Apple, Alphabet, Amazon, Facebook, and Microsoft.

Monopolies in the twenty-first century—the platforms that are becoming the highways and power plants of the unscaled economy—aren't going to look like the monopolies of the past. They will quickly become global, making it difficult for any one nation to regulate them. And there might not be any easy way to break them up or undo them because their AI—which gets so good because of scale—will become so important to so many businesses that use the platform. Break up the company, and the AI won't be as good and the companies that use the platform will suffer.

Blockchain technology might bring us an alternative to some of these platforms. Instead of a platform belonging to a company, a platform could belong to no one—along the lines of open-source software or Wikipedia. Blockchain technology runs on thousands or millions of independent computers spread around the globe, according to rules built into the software. No monopoly controls it. Bitcoin—a currency controlled by no central bank—is built on blockchain. So imagine services or applications built on blockchain that no company can dominate. One company, Blockstack, is developing a blockchain applications platform that aims to make it easier for developers to build blockchain-based apps and offer them to the public—something of an Apple App Store for blockchain, but with the community in charge instead of a company. If some of these open platforms catch on, they might create competition for monolithic commercial platforms, the way open-source operating system Linux ate into Microsoft's near monopoly in the late 1990s. So blockchain might be able to give us the benefits of platforms without creating powerful tech companies that gain too much control.

Our policymakers need to understand and debate the monopoly platform issue. The economy will need these platforms to unscale, yet the platforms must enable the creation of free markets or risk getting regulated.

ALGORITHMIC ACCOUNTABILITY

The 2016 US presidential campaign raised awareness of another new threat that the technologies of unscale could unleash: AI algorithms that behave badly. We saw it in the spread of "fake news" on Facebook. The site's AI is tuned to serve users content that will most likely spur them to continue to interact with the site—in turn giving Facebook more opportunities to serve up ads and make money. Inflammatory made-up news that plays to people's biases tends to engage us, so Facebook's AI learns that lesson and serves up more of the same. The algorithm doesn't know real news from fake, only which content serves Facebook's purpose. Although no one can be certain that such fake news affected the election results, it did expose how easily algorithms can be gamed. It's a very dangerous phenomenon as we enter an era when AI will make all kinds of decisions for us.

With AI algorithms, companies can optimize for every variable. Uber efficiently dispatches its drivers to high-traffic areas during rush hour, and Amazon can suggest the right product at the optimal time to get someone to make a purchase. For all their positives, algorithms aren't optimized for doing the *right* thing or displaying any amount of transparency. The algorithms that manage internal operations at companies strongly favor the most expedient, efficient, and effective practices. If that means excluding, say, certain minorities from getting services, the algorithm doesn't care. It will only care if humans program it to care.

Concern about "black box" algorithms that govern our lives has been spreading. New York University's Information Law Institute hosted a conference on algorithmic accountability, and Yale Law School's Information Society Project is studying this too. "Algorithmic modeling may be biased or limited, and the uses of algorithms are still opaque in many critical sectors," the group concluded.

This needs to change, and companies must take the lead in creating algorithmic accountability in their services. The industry shouldn't rely on government to play this role. Old-school regulations written by lawmakers and bureaucrats are too cumbersome to keep up with technology, and government-led regulation would burden tech companies and slow innovation. To avoid that fate, these new utilities—the Googles, Amazons, and Facebooks of the world—must proactively build algorithmic accountability into their systems and faithfully and transparently act as their own watchdogs. In fact, we need a new version of the types of standards-enforcing bodies that emerged to oversee the new technologies of the last explosion of invention in the early 1900s. As complex electricity-driven innovation entered the market, in 1918 five engineering societies and three government agencies joined to found the American Engineering Standards Committee, which eventually morphed into today's American National Standards Institute (ANSI). The role of the organization has always been to ensure transparency (so users know what they're using) and safety and, by doing so, keep lawmakers from imposing government oversight.

In similar fashion today's AI companies might create their own standards and transparency requirements. Watchdog algorithms that monitor companies' algorithms could even work like open-source software—open to examination by anyone. That way, coders could see whether the watchdog algorithms are monitoring the right things, while the companies keep proprietary algorithms and data to themselves. Technology companies and policymakers need to come together and share ideas. It's clear that if companies follow the current path of little algorithmic accountability, the government will institute regulations to oversee social networks, search, and other key services.

* * *

Every industry will undergo radical change in this era of AI and unscaling, and those changes will raise significant questions for policymakers. Although it would be impossible to detail everything we need to be thinking about as a society, here are some of the more important issues I see coming.

HEALTHCARE AND GENOMICS

The medical sector will raise some particularly difficult issues, especially around genomics. Inventions like CRISPR allow us to edit genes and, thus, alter people. We're close to being in control of our own evolution. I can foresee a startup eventually offering gene editing that lets customers upgrade their bodies or brains. If that comes to pass, we risk opening a biological divide that's far more damaging than the old digital divide. Wealthy people will have the opportunity to make themselves better, healthier, and smarter than poorer people, creating a gap between rich and poor that's not just about wealth and opportunity but about talent and physical prowess. It's bad enough when societal barriers make it difficult for people to improve their lives and rise through the economic strata; it's a whole new problem if a class of people are denied the physical and mental ability to compete with those who also have a financial advantage. We could wind up with a society of permanent classes. Governments must think this through before it becomes a reality.

Genomics startups are already getting ahead of laws. Consumer genetic testing company 23andMe ran into just such trouble. In its early years (the company was founded in 2006) 23andMe offered consumers a DNA test that could predict a person's propensity to develop certain conditions, from baldness to Alzheimer's. In 2013 the US Food and Drug Administration (FDA) ordered it to stop, fearing that the company's interpretations of the tests could mislead consumers about their health prospects and cause harm. 23andMe continued to offer tests to trace ancestry but stopped the predictive interpretations. The FDA wanted time to study genetic testing products. In April 2017 the FDA finally gave 23andMe approval to run tests that can tell a consumer if they're at a higher risk for ten diseases, including Alzheimer's, Parkinson's, and celiac.

The 23andMe case is just an early sign of the policy debates to come. The National Human Genome Research Institute calls for a range of issues to be studied and debated. Could employers demand your genetic data and use it as a basis for hiring? Could insurance companies use it to decide how much to charge you because you might use a lot of healthcare resources if you're predisposed to a disease like Parkinson's? If you get a genetic test done, who owns that data? The testing company? You? To truly unscale healthcare and create predictive medicine, your genetic information will need to be shareable with any doctor, hospital, or application you're using—yet genetic information is so sensitive that you'd want to be able to tightly control it and give it only to those you choose. How do we do that? "Genomic research alone is not enough to apply this new knowledge to improving human health," the Institute writes. "We need to carefully study the many ethical, legal and social issues raised by this research."

The 23andMe case shows that the FDA might need to change its mission in an era of radical change in medicine. The FDA's approval process is just too slow—it takes an average of around ten years of studies to bring a new drug to market. The FDA process is designed to determine both safety and efficacy: Will the drug do unintentional harm, and will it do what it's intentionally supposed to do? A growing number of voices in medical technology argue that the FDA should mainly regulate for safety and leave efficacy to the market and to data analysis to solve. In other words, the FDA should be in the business of making sure a drug, test, or procedure doesn't do irreparable damage. Then put it on the market. Now that we're collecting so much data about patients, we'll be able to quickly see how a drug, test, or procedure is performing—much more quickly than from a series of small studies. On top of that, in this era of social media and Yelp-like ratings, complaints about a drug not working would surface quickly. This kind of thinking is reflected in an FDA reform called the Free to Choose Medicine proposal, which groups such as the Mercatus Center at George Mason University and the Federalist Society have discussed. The proposal essentially says the FDA should still go about its approval process but also have a second track for drugs proven safe but not proven effective to allow doctors and patients to choose to try those drugs.

Another upcoming issue in healthcare will involve access to our personal health data. Over the past decade medical records—long the realm of paper and doctors' horrible handwriting—have been digitized and fed into software. That data is trapped in private systems that don't get high grades for usability or easy access. Epic Systems, the biggest electronic health record (EHR) company, supplies EHR software that holds 54 percent of US patients' records despite getting bad marks for usability flaws that eat up doctors' and nurses' time. One report from *Becker's Hospital Review* said that almost 30 percent of Epic clients wouldn't recommend it to their peers. Epic and competing systems are closed, making it difficult for doctors and hospitals to share data. What's more, consumers have little control over that data—we can't easily get it and mash it up with other data about ourselves that might, in the aggregate, help us stay healthier. All this holds back unscaling of healthcare by locking out upstart companies that could use the data to offer innovative medical care. I believe these companies should be forced to give patients more control over their own data and allow aggregated (and anonymized) data to be used for research and learning about health and medicine.

As healthcare unscales, soon the partisan debate about the Affordable Care Act (also called Obamacare) will seem silly. Startups and floods of data and automation promise to drive down the costs of healthcare while increasing effectiveness. If this trend plays out, within five to ten years Congress won't need to fight about the exploding costs of Medicaid and insurance—it instead might battle over what to do with a leftover windfall. Consider the example of Livongo from earlier in the book: their AI-driven technology is designed to help keep people with diabetes healthy, resulting in fewer trips to emergency rooms and fewer expensive complications. From what I've seen in the field, I believe technology can easily take $100 billion out of the annual cost of treating people with diabetes in the United States. Imagine applying that to many other chronic conditions. Congress needs to start considering what healthcare for all might look like if costs are falling and healthcare delivery through apps and devices becomes widespread.

ENERGY

AI is enabling the world to inexorably move away from carbon and toward solar and other renewable and unscaled sources of energy. In his first months in office President Trump signed orders to try to save coal jobs, eliminated rules that pushed automakers to develop electric cars, and put in place an energy secretary who believes climate change isn't real. All this could encourage US companies to stop investing in next-generation energy. Meanwhile the Chinese government has a policy to pump billions of dollars into solar, battery, smart grid, and electric-vehicle companies. This is where policy decisions made today will have an enormous impact on how unscaling plays out in the next two decades. Current US policies seem to be a formula for falling behind, while Chinese policies are taking advantage of the forces of the AI century.

Governments need to recognize how energy is changing and to help innovators to benefit society and business. For instance, the transition from gas-powered vehicles to electric vehicles will be an enormous change. How do we switch over hundreds of thousands of gas stations? How do we build a power grid that can handle all that new demand? What happens to state car inspections and Jiffy Lube franchises? Our policy debates need to start with the obvious endgame—electric cars and trucks are going to disrupt gas cars and trucks—and then ask how to best make that transition.

Similarly we know that ultimately solar and renewable power will replace carbon power, so how do we guide that change? We know that electric power will unscale into many small producers that will want to sell power back into the grid, so how do we develop an open, two-way power grid? How do we help utilities make themselves into platforms that support thousands or millions of small power producers?

As we've seen in recent decades, energy companies, especially those in oil, are entrenched and wealthy, spending enormous amounts of money on politics and lobbying efforts to defend their business interests by holding off the transition from carbon. We need to help everyone see that getting off carbon isn't a negative—it's actually the world's biggest economic opportunity and a huge job-creation engine. Imagine all the work needed to replace all the old gas stations, install solar panels and batteries in every home, and rebuild the old power company infrastructure with networked two-way grids. Those aren't jobs that can be outsourced to another country—they need to be done on the ground locally. If embraced, this great energy transformation could help put people to work while AI automates other jobs away.

FINANCE

Laws and regulations that govern banking were set up for an era of financial giants and concerns about banks becoming "too big to fail." But as finance unscales, policy will need to take a different tone. Big banks won't be the challenge—policy will need to address what to do about a flood of upstart, software-driven financial offerings that might only exist in an app.

Companies like Digit, as described earlier, are evolving into financial services that sit on top of big banks, using those banks as a platform. Policy needs to set up rules for this kind of transition, allowing banks to unbundle themselves and become platforms for unscaled newcomers. How can consumers be protected in such a scenario? Does the FDIC's deposit insurance extend to apps? Should big-bank platforms handle all banking's compliance and legal issues so the startups don't have to—or if a startup acts like a bank, should it be regulated like a bank?

AI and unscaling lead to a discussion we should be having about the Federal Reserve. Instead of collecting and analyzing old data and then meeting eight times per year to decide whether the federal funds rate needs to be adjusted, the Fed could rely on AI to analyze data as it happens—data from unemployment offices, the stock market, retailers like Walmart, and shippers like FedEx. As the AI learns about the economy, it could constantly suggest ways to react just a little ahead of conditions. Instead of adjusting interest rates every few months based on past data, the Fed might continuously adjust rates—not by quarter percentage points but by tiny increments as small as 1/100th of a point—based on what's about to happen. In other words, it will become possible to build AI-driven financial policy. Do we want that? How would it work? Again, these are questions policymakers need to consider today so they can shape our tomorrow.

EDUCATION

With the right policies, unscaling can help us create better K-12 schools that fit our times and solve the problem of making college affordable. It might even make traditional four-year college unnecessary for many careers.

Creating better K-12 schools has been a constant goal for decades. We try charter schools and vouchers, team teaching, and Common Core. All these have been top-down, scaled approaches. Policymakers should instead find ways to let teachers improve their individual classrooms so each classroom becomes its own focused unscaled unit built around the needs of the particular students in that class. Help teachers adopt new technologies that allow them to connect their classroom to parents and to other students and then integrate coursework from wherever it might be generated in the world. Let teachers rely on

AI-driven coursework—like the classes from Khan Academy—that helps each student learn at their own pace so the teacher can be a coach instead of droning through a state-written curriculum.

The costs of college are unsustainable, burdening new graduates with mountains of debt. College courses online—some from newcomers like Khan or from established schools such as MIT and Stanford—offer an alternative path. But that would require new approaches to credentialing colleges. The only way online courses can take the place of spending four years on campus is if you can get a credentialed degree online. Employers can drive this change around credentials on their own by putting less emphasis on prospective employees having a credentialed degree and instead accepting that online work might add up to an education that's just as good—or maybe even better. We see some forward-thinking companies such as Google placing less emphasis on formal education when hiring. If that becomes more widespread, more people will consider alternative ways of learning.

The bottom line is that education systems need to match economic systems. In the previous century we built scaled-up schools to match that era's scaled-up factories and corporations. The education system churned out people suited for the structure of that economy. Today's schools are outdated because while the economy is unscaling and changing, most schools still function as they did a hundred years ago. How we teach is unsuited for the times. Over the next decade policymakers and educators will need to fix this and align education with the economy. There's just no way around it.

10

The Corporation

Charting an Unscaled Future for Scaled Enterprises

In 1837, more than two decades before the American Civil War, William Procter and his brother-in-law, James Gamble, formed a company in Cincinnati, Ohio, to make candles and soap. Cincinnati at the time was home to a booming hog butchering trade, and candles and soap were both made from a byproduct of that trade, animal fat. The cofounders called their enterprise, simply enough, Procter & Gamble.

The company grew slowly and got a boost from contracts with the Union Army during the Civil War. Its breakthrough came in 1878, just as newspapers were reaching consumers en masse and railroads opened that could efficiently carry products to any major city. According to lore, one of the company's chemists accidentally left a soap mixer on during lunch, stirring more air than usual into P&G's white soap. The air made the soap float. The company branded the product as Ivory and marketed the floating soap nationwide. P&G began to scale up. By 1890 it was selling thirty different kinds of soap. In 1911 it introduced Crisco shortening, moving into food. After World War II, as the consumer market took off, P&G brought out Tide detergent, the first mass-market soap specifically for cleaning clothes in an automatic washing machine. By the end of the twentieth century P&G had scaled up to a behemoth, offering more than three hundred brands and raking in yearly revenue of $37 billion. P&G was one of the world's corporate superpowers.

In 2016 analyst firm CB Insights published a graphic showing all the ways unscaled companies were attacking P&G. It looks like a swarm of bees taking down a bear. In that rendering P&G no longer appears to be a monolithic scaled-up company that has built up powerful defenses against upstarts; instead, it is depicted as a series of individual products, each vulnerable to small, unscaled, agile, AI-driven, product-focused, entrepreneurial companies. P&G's Gillette razors were being successfully challenged by Dollar Shave Club and Harry's newfangled subscription models; a niche of buyers of P&G's huge Pampers brand of disposable diapers were getting peeled off by The Honest Company's environmentally friendly diapers; Thinx "period panties" were going after P&G's Tampax tampons in a new, uncharted way; and eSalon "custom" hair coloring was challenging P&G's Clairol mass-appeal hair coloring.

CBI called the overall phenomenon the "unbundling of P&G." It is as clear an indication as any of what big corporations face in an era that favors economies of unscale over economies of scale. Small unscaled companies can challenge every piece of a big company, often with products or services more perfectly targeted to a certain kind of buyer—products that can win against mass-appeal offerings. If unscaled competitors can lure away enough customers, economies of scale will work against the incumbents as fewer units move through expensive, large-scale factories and distribution systems—a cost burden not borne by unscaled companies.

But if that's the new reality, what's a corporation to do in the AI century? How can a company built to take advantage of scale reverse course and take advantage of unscale?

It won't be easy, but some forward-thinking companies recognize what's happening and are experimenting with responses. One of them happens to be P&G. For about a decade it's been running a program called Connect + Develop. After 175 years of inventing most of its new products in house, the company's executives came to understand that there were more smart inventors outside of P&G than could possibly be contained inside P&G, and the internet provided a way for outside inventors to connect to P&G. The program invites anyone who has developed a product that would be a good fit with P&G to submit proposals. Though P&G never phrased it this way, Connect + Develop is essentially a way for the company to become a platform for niche products in a way that benefits P&G (it gets to capture some of the value of new unscaled products instead of competing against those products) and benefits product innovators (who can "rent" P&G's distribution, marketing, and knowledge to get a product to market).

Connect + Develop hasn't completely transformed P&G from a scaled company to a new model of unscaled company, but it has moved P&G down the right path. According to one 2015 study, about 45 percent of initiatives in the company's product development portfolio had key elements discovered through Connect + Develop. A future unscaled version of P&G might look more like a giant consumer products platform that a constantly evolving swarm of small, focused entities rent—an Amazon Web Services model for tangible consumer goods.

General Electric is another old, enduring company trying to stay vital in the unfolding unscaled era. GE's big bet is on an AI platform called Predix. For most of its history GE has built industrial products—complex machinery that undergirds much of business. It builds train locomotives, airplane engines, factory automation machinery, lighting systems, and so on. In the 2010s GE pushed hard into IoT, rightly understanding that many of its industrial products were already jammed with sensors designed to measure the machines' efficiency and well-being—and with IoT those sensors could communicate their data back through the cloud to AI that could use that data to learn even more about machines in aggregate.

The streams of data from Predix can help GE optimize its products for its customers. What AI learns from all GE locomotives can help a railroad company better operate its particular GE locomotives. In this age of unscale GE also opened up Predix so others could build on top of it. The company calls Predix "a cloud-based operating system for industrial applications." Other companies can use it to create software that learns how to better run factories. GE has a Predix Catalog that operates something like an app store for industrial developers. As the Predix blog says, "The catalog consists of more than 50 services and analytics designed to save you time and effort while meeting the requirements for the Industrial Internet of Things. And, whenever you create something that can be reused by others, contribute it back to the catalog for your organization (or even to GE Predix!)." The company even hosts a conference called Predix Transform, where industrial developers learn from one another and help build a Predix ecosystem.

As with P&G, Predix by itself isn't overhauling GE, but it is a way for GE to take advantage of unscaling by using its skill set and data to create a platform that others can rent.

In 2016 Walmart spent $3 billion to buy Jet.com. (Our firm was an investor in Jet.) As discussed in Chapter 8, Walmart was a superstar of scaling and now is supremely vulnerable as retail unscales—which is why it would pay $3 billion to buy a barely proven company. Jet.com is an AI platform for other retailers. It uses sophisticated AI to constantly adjust prices depending on many factors, including how much the customer is ordering right then and how far the customer is from the product. The goal is to give consumers the lowest prices possible—even lower than Walmart's. Most of the products come from independent retailers—more than two thousand sell on Jet. The pitch to retailers is that Jet itself won't compete against the retailers, unlike the way Amazon can compete against retailers who sell through Amazon Marketplace.

So through one lens Walmart bought Jet to get its brain trust and innovative technology. Through an unscaling lens, though, it looks like Walmart is trying out a platform strategy. Perhaps Jet will evolve into a way for focused, niche consumer retailers to rent the power of Walmart's platform to sell physical products to anyone anywhere.

* * *

Over the past hundred years, as the era of scale unfolded, small companies of course continued to exist, and many prospered even as they stayed small. Small business was the US economy's underlying strength throughout the scaling age. In 2010, according to the US Census, the nation had about 30 million small businesses and only 18,500 companies that employed more than five hundred people.

However, in an era when economies of scale usually prevailed, when a scaled-up company competed directly against a small business, the small business usually lost. Just think of all the small-town Main Street retailers Walmart bulldozed over the past twenty-five years.

We will see the big-beats-small dynamic reverse as we unscale. Over the next ten to twenty years companies that relied on scale as a competitive advantage will increasingly find themselves defanged. They will be at a disadvantage against focused unscaled businesses. Large corporations won't disappear, just as small business didn't disappear in the last era. But the big companies that don't change their model will see their businesses erode, and some of today's giants will fall.

The best *Fortune* 500 leaders will find ways to reinvent the corporation for the era of unscale. It's hard to know exactly how that will play out, but we can see the outlines of some of the tactics emerging today—like those of P&G, GE, and Walmart. Here are a few ways corporations can stay relevant and play important roles in an unscaled economy:

BECOME A PLATFORM: This is the way Connect + Develop, Predix, and Jet lean. In Chapter 3 I discussed how electric utilities need to morph toward a platform mindset and make the grid into a system that can support thousands of small energy producers. In Chapter 6 I outlined how major banks can become platforms for small, focused financial apps like Digit.

That's not to say every corporation must become a platform or perish. But a successful platform strategy looks like a path to growth in the unscaled era. Platforms can be enormously profitable and enduring because a whole ecosystem of companies comes to depend on the platform for *their* success. This is why the AWS cloud platform has become the profit engine for Amazon, with operating margins of 23.5 percent, compared to about 3.5 percent for the Amazon retail business.

Vibrant corporations have spent decades building scale that's highly specialized for their industry. They've built efficient factories, distribution channels, retail outlets, supply chains, marketing expertise, and global partnerships. But now they need to ask themselves if there's a business in simply and elegantly renting that capability to others, much as AWS rents computing capability to more than a million active customers.

Imagine Ford as a car-making platform that allows hundreds of small companies to design innovative new vehicles and get them made, marketed, and delivered to customers—all in a way that allows these small car makers to serve a niche market at a profit. No doubt we'd get some pretty interesting new cars on the road. Or maybe Anheuser-Busch InBev stops buying beer brands and instead becomes a beer platform, allowing microbrews to rent its capabilities to get new beer concoctions to market with a few clicks on a web page.

This is somewhat of an inverse version of how technology companies have become platforms over the past few decades. Tech startups usually start out doing something small, like a single-purpose application. Facebook began as a student directory at top universities. Stripe started by just processing payments. As companies get to a critical mass of customers, the smart ones start allowing outside developers and users to build on top of their technology. Facebook lets users create pages for businesses and rock bands as well as allowed game developers and news media to publish on its platform. Stripe created the Connect platform and then launched it by putting up a $10 million fund to support companies that build on it, and it has since launched Atlas, a service that takes care of almost everything a startup in any part of the world needs to get up and running as a US-incorporated company. "Over time we want to increasingly manage all the business and revenue things so a startup can focus on building its products and on what differentiates it," says cofounder Patrick Collison—a big, broad goal that would position Stripe perfectly to both take advantage of and drive the unscaling era.

So tech companies engineer a platform from the bottom up. Big corporations will need to pull off something quite new: creating platforms by taking themselves apart.

By the way, in the platform business one company usually takes most of the market. Just look, as noted earlier, at Microsoft's and Google's difficulty competing against AWS. So in any one market sector the corporation that transitions to a platform first might hold a distinct advantage. If I were a *Fortune* 500 CEO, I'd be studying a platform move right now and investing in early versions like Connect + Develop, Predix, and Jet.

RADICALLY FOCUS ON PRODUCT: As companies get big, the focus often gets lost amid process, bureaucracy, politics, concerns about stock price, and a whole lot of other stuff that has nothing to do with making a great product for the exact people who need that product. Big corporations try to create products that appeal to the most people possible so they can achieve economies of scale and become more profitable. But in an unscaled era making such mass-appeal products becomes an Achilles heel—a set-up for a product-focused small competitor to knock down.

Big companies in the unscaled era will look more like a network of small entities, each absolutely committed to making a product that's perfect for its slice of the market. Everything else a corporation does will wind up being rented. Outsourcing has been a trend for several decades—companies have been shedding "noncore" tasks, which is why an Apple or Nike contracts out manufacturing to China while Netflix runs its streaming service on AWS instead of building data centers. But next-generation unscaled corporations will outsource far more. Payroll and other human resources functions could be rented from a company like Gusto, which today serves small businesses but tomorrow could handle big companies. Payments processing can be rented from the likes of Stripe. Anything that doesn't have to do with developing a great product needs to go.

A new kind of management team would lead such a product-intensive corporation. Instead of MBAs trained in the process of business, product and platform people will lead successful corporations. Perhaps the best recent example of this was Steve Jobs's Apple. Apple historically wasn't an unscaled kind of company—it was very much a mass-market, you'll-like-what-we-say-you'll-like company. But Jobs-era Apple became a company that created platforms, especially the iPhone, App Store, and iTunes. So that type of shift at a big company can be done.

An unscaling mentality affects almost everything a corporation does. It changes who the company hires—product people, not process people—and what kinds of investors buy its stock. It takes the emphasis away from brand—because brands are an artifact of mass-market consumer culture—and puts it on experience. It flips an org chart from top down to bottom up—the product creators will drive decisions while top management provides the platform for them to build on. The *Fortune* 500 corporation of twenty years from now is likely to be smaller, faster moving, and more like a network of small companies than the business giants of the early years of this century.

GROW BY DYNAMICALLY REBUNDLING: The winners in the unscaled economy will make every customer feel like a market of one. Products and services that can be tailored to the individual will beat out mass-market products and services. But I can foresee one way a corporation—that is, a collection of products—can have an advantage. Once it comes to understand a particular customer of one of its products, the company can offer that customer other products from its portfolio that it knows fit them. A big company could, in fact, bundle together products tailored to each customer.

To get a sense of how this would work, take a look at how The Honest Company grew. By 2012 Honest started selling a line of organic diapers and wipes by subscription. That first year the company pulled in $10 million in revenue. Honest had won over a certain kind of niche customer who wanted a certain kind of niche product different from mass-market brands. The company used that knowledge to develop other products in the same vein—shampoo, toothpaste, vitamins. By 2016 it had 135 narrowly focused products. Honest could then bundle the right set of products for the right customers, making those customers feel like a market of one for Honest. In 2016 sales passed $150 million. In a way, Honest had become a mini-P&G, offering a variety of items—but with a big difference. Honest knew its customers and could bundle its various products accordingly. Each of P&G's products is a stand-alone brand, sold in stores to people P&G will never know or understand the way Honest does.

This kind of rebundling allows a company to mimic the advantages of scale without actually building scale. The company can stay nimble and innovative, focusing on product and using its portfolio to expand its sales to each individual customer. So a P&G of the future might operate as a platform for thousands of product-focused entities yet be intelligent enough to understand each customer and dynamically rebundle a set of those products for every individual. This, it seems to me, is how the smartest corporations can play in the unscaled age.

* * *

If you want a glimpse of the future corporation, look first at the ethos behind Amazon—and then keep an eye on an experiment, called All Turtles, that my firm is funding.

In early 2017 Amazon CEO Jeff Bezos published a letter to share-holders about a concept he calls "Day 1." As he wrote, "I've been reminding people that it's Day 1 for a couple of decades"—which is nearly the entire time Amazon has been in existence. "I work in an Amazon building named Day 1, and when I moved buildings, I took the name with me. I spend time thinking about this topic."

Amazon is a huge company. In the fourth quarter of 2016 alone it pulled in $43.7 billion in revenue. At the time Bezos published his letter Amazon was worth about $434 billion, making it the fourth most valuable company in the United States, behind Apple, Alphabet (Google), and Microsoft. Yet Bezos has strived to make sure Amazon is always a very unscaled giant. Day 1, Bezos wrote, is about constantly creating new, nimble, product-focused businesses inside Amazon—businesses that can be quickly built on top of Amazon's corporate platform and feel like they are in the first day of life. To Bezos, Day 2 is when a business gets bogged down by its own scale.

How does Bezos manage this? "I don't know the whole answer," he admits in the letter, "but I may know bits of it." He offered four points that he considers a "starter pack of essentials for Day 1 defense." They align with what we know about unscaling.

The first is "true customer obsession." In this unscaled era the products that win make you feel like a market of one. Doing that requires deep knowledge of the customer and a willingness to build products that perfectly address a certain segment, however small. Big companies usually fail at this. They strive instead to build products for the broadest possible set of customers. "Staying in Day 1 requires you to experiment patiently, accept failures, plant seeds, protect saplings, and double down when you see customer delight," Bezos wrote. And because of that approach, over the years Amazon has brought us things like the Kindle, Amazon Web Services, and Alexa. The company seems to renew itself constantly.

Bezos's second tactic is "resist proxies." Scaled companies can get lost managing things that don't matter. One example is process. To manage a sprawling empire, companies create processes for employees to follow. Too often, Bezos writes, "the process becomes the thing. You stop looking at outcomes and just make sure you're doing the process right." Other bad proxies include market research in place of actually knowing customers. "You, the product or service owner, must understand the customer, have a vision, and love the offering." That intentionally sounds like instructions for a startup company. Bezos wants Amazon to feel like a collection of startups.

His third point: "embrace external trends." As Bezos notes, "The big trends are not that hard to spot (they get talked and written about a lot), but they can be strangely hard for large organizations to embrace." Newspaper companies saw the internet coming from a mile away, for example, but delayed moving online until it was too late. If a big company operates as a collection of nimble small companies, the small companies are more likely to spot and react to new technologies as tastes shift.

The final Day 1 point is about "high-velocity decision making." It fits right into the unscaling playbook. As Bezos writes, "Never use a one-size-fits-all decision-making process." Let the smaller units make their own decisions based on their insight and their customers' realities. The more a company scales, the more complex it gets, and so decisions seem complex. Executives feel they need a ton of input and information before making a decision. All that leads to stagnation and the onset of Day 2. Companies need to make decisions like it's Day 1 and then move on if the decision proves to be wrong.

As Bezos admits, these concepts probably only begin to describe how a giant company can operate as an unscaled organization. But Bezos certainly believes that today's corporations *must* operate as unscaled organizations.

This brings me to All Turtles. It is an experiment our firm launched in the spring of 2017, and we don't know how it will turn out. But it is a bet on the future of entrepreneurship in an unscaled era as well as on many of the concepts in this book.

All Turtles is the brainchild of Phil Libin, who joined General Catalyst after working as CEO of Evernote for eight years. Around the firm we talked often about unscaling and what it might mean to the way businesses are created and structured and how that might impact the VC model. If you take unscaling to its logical conclusion, you might wonder why anyone would start a *company* at all. Someone with an innovative product idea should be able to essentially *rent* a company; in other words, an innovator might let someone else assemble all the scale—computing, cloud, finance, payments, engineering, marketing, distribution, legal, and so on—into a rentable platform. All the innovator would need to do is build the product and plug into the platform for the rest.

But, we wondered, is there a way to do that and still allow innovators to benefit from the collective of businesses on the platform? We were looking for some optimal unscaled blend of the way VC works and the ideas behind Amazon's Day 1 doctrine.

Phil came up with what he calls a studio model. We think of it like modern TV studios run by HBO or Netflix. Someone comes to the studio with a good idea. If the studio likes it, it can say, "Okay, we'll fund the pilot and assemble the professionals to make it. If we like the pilot, we have a platform—a studio and distribution system—for making the series and putting it in front of an audience." As the idea person, you don't need to hire a team, raise money, find office space, or any of that—you just need to create and let the platform do the rest.

But we wanted All Turtles to be more than that too. These studios will be physical spaces, where our innovator-entrepreneurs can work and cross-pollinate. Libin plans to put these studios all over the world—in fact, we hope it will be a way to harness good ideas from the rest of the world that's *not* Silicon Valley. Further, we want all our innovator-entrepreneurs to have a stake in the whole ecosystem. So if All Turtles brings you aboard, you primarily have equity in your venture, but you also get some equity in the whole pool of projects—a motivation to help others in the All Turtles collective.

"Call it a fellowship," Phil says. "If you're in it, you get a say in who else can be in it. We think it can disrupt the idea of a company with a more honest structure. You become loyal to the fellowship. The goal is to get the best people to make great products and dramatically expand who gets to work on stuff and what gets funded."

If it works, in a couple of decades All Turtles could become a gigantic global enterprise that is, at the same time, entirely unscaled. It will then be the uncorporation—a new blend of product-focused small companies addressing highly targeted customers with the platform-like aspects of a world-spanning corporate giant.

Remember, the corporation hasn't been around since Adam and Eve. It was an invention of the Industrial Age, created to manage scale and complexity under one roof. The corporation both enabled scale and is a product of scale. It makes sense that the era of unscale needs a new structure. Maybe it will look like All Turtles, and maybe it will look like something else that doesn't yet exist. But surely some kind of *un*corporation will emerge in the near future.

11

The Individual

Living Your Life as
a Personal Enterprise

I have three children. As I'm writing this chapter they are aged thirteen, eight, and three. The transition to an unscaled economy will powerfully impact their lives. So what do I tell them about education, careers, and life? Their paths might be completely different from what I knew. For instance, given my kids' ages, I think the eldest will find that going to a traditional four-year college is still the best way to move into the working world—but I'm not sure that will still be the case for my youngest. By the time they reach college age, going to a traditional college may no longer be the best option.

The way we all think about our education will need to change because the way we think about our work will need to change. They go hand in hand. In fact, one of the truths about the unscaled era is that work and education will blend. The idea that we go to school for twenty or so years and then work for the rest of our lives will seem silly. Instead, we'll learn and work throughout our lives—we'll start working earlier in life and keep learning brand-new things much later in life.

So how should an individual plan to navigate the unscaled economy? The key is to live what I call an *entrepreneurial life*.

Most people in the twentieth century did not live an entrepreneur-ial life. Of course there have always been entrepreneurs, even before we called them entrepreneurs. But in the scaled economy it was common to think in terms of getting a job and having a career. For Americans of previous generations, the path to success was good education, a full-time job with a big corporation, a career path that would take you up the corporate ladder, and retirement at sixty-five with a pension. By the time much of today's workforce came of age, that lifelong strategy was already becoming unhinged. In the next couple of decades it's going to break apart completely.

Keep in mind that the full-time job is not some natural state of human existence. Before the mid-1800s few people worked a struc-tured work week. Early industrialists dreamed up that concept because they needed to bring workers together in a factory at the same time to efficiently make products or into an office at the same time because the only way to collaborate was to physically sit in the same room. For the past hundred years the forty-hour-a-week job has been the centerpiece of work life because there was no better way for people to gather in one place at the same time to get something done.

Big companies kept employees for decades as part of building scale. If they were scaling up, constantly adding pieces to the company and building higher barriers to entry, they needed people—lots of people. Companies in the past would keep the people they had and add more. Getting bigger was the point.

In an unscaled economy staying close to a startup mentality—or, in Jeff Bezos's formulation, staying close to Day 1—is a better strategy. Better to keep the core workforce small and rent the rest, including renting skills and labor through sites like Upwork or labor unions as well as renting capabilities that can be automated by software from companies like Gusto and Stripe. The internet, cloud computing, soft-ware, 3D printing, and other new technologies enable people and small companies from all over the world to collaborate and get something done without ever being in the same place.

Big factories and offices will give way to dispersed work in the cloud, with jobs always shifting and morphing, depending on what's needed at the time.

For many people the twentieth-century-style, full-time job will disappear. And the concept of a career that steps along a defined path in a single field will wane with it. We're already feeling all this in our society. We see it in factories that lay off workers because of automation and outsourcing or in the rise of the gig economy as people piece together new ways to make a living. Just as unscaling is unbundling the company, it is also unbundling work. The job market becomes a market-of-one, just like everything else. An employer doesn't always need all your skills or time—it usually really just needs a specific thing you can do for a specific period of time.

The trend toward unscaling and unbundling work is only going to speed up. No politician is going to stop it, though some might slow it down for a bit. As an individual in the emerging unscaled economy, your best bet is not to fight the oncoming forces but to take advantage of them. Although unscaling is eroding the old idea of job security, new opportunities are coming in its place.

Over and over in this book you've read that unscaling is making it easier for anyone to start a business and rent scale to compete. That means that anyone with an idea can quickly and inexpensively get into business. So first and foremost, successful people in the unscaled age will be entrepreneurial. They won't all be the next Mark Zuckerberg, starting a company that becomes all-consuming for years; many will start multiple small enterprises that come and go throughout a career. This idea that we will live more entrepreneurial lives has become accepted wisdom in tech circles. In 2012 LinkedIn cofounder Reid Hoffman and Ben Casnocha published *The Start-Up of You*. "All humans are entrepreneurs," they wrote in their book. To accelerate your career in today's economy, they say, you need to embrace that spirit. Many books and self-help seminars these days are expressing that same sentiment—for good reason. It's the way to do well in an unscaled economy.

Being an entrepreneur doesn't only mean starting a company. Technology platforms are creating all sorts of ways to sell yourself and your assets. We already see how that works with, for instance, Airbnb, which lets people make a business out of renting a spare room, or a car-sharing platform like Getaround, which lets people rent out their personal cars. Uber, of course, gives people a way to make money by driving. Platforms like Upwork allow individuals to market their skills as writers or coders. Shapeways allows anyone to design a product and have it 3D printed and sold worldwide. All these make it possible for a typical professional to have a multifaceted career with several sources of income.

Is this good or bad? It depends on how you view it. A lot of people will lament the loss of security, continuity, company-paid benefits, and other artifacts of the corporate job. For a lot of people, though, getting a good corporate job was difficult in the first place. And a lot of employees who had corporate jobs found them deadening—stuck doing something from 9 a.m. to 5 p.m. every day that, to say the least, did not ignite their passions.

As work unscales, each of us will have more choice about what to work on and when. The idea is to double down on your passions. Find the things you really want to do—the things you're especially good at—and market them to all comers. Unscaling offers us the opportunity to do what we really love. It puts each of us in charge of our work—and transfers more of the responsibility for our well-being and our incomes onto us. It's a burden to bear, yes—and a freedom to enjoy.

Although this kind of dynamic, entrepreneurial life may feel painful to an older generation, we already know that younger workers tend to prefer their work to be more unscaled. A recent Future Workplace survey found that 91 percent of millennials expect to stay in a job for less than three years and that they want flexible hours and the ability to work wherever they want. Those flexible policies are more important to the young generation than salary.

Will full-time jobs evaporate completely? No. But such jobs will be different from the past. In an unscaled economy, as Gusto CEO Joshua Reeves says, small companies can compete against corporations for the best employees—just as unscaling allows them to compete against giants for customers. So more high-performing people who desire full-time work will find such jobs at small, unscaled companies instead of big corporations.

Taking a job at those small companies will feel more like joining a community than plugging into a corporate hierarchy. The community will expect you to contribute, but you won't be ordered to. Responsibility for your success will be more on your shoulders and less on the company's. In other words, even in a "real job," you'll need to be entrepreneurial to do well.

Living an entrepreneurial life is also a way to stay ahead of artificial intelligence. AI will increasingly automate routine work—even the work of some kinds of highly paid professionals. Wall Street traders and radiologists are in as much danger of finding their jobs automated away as truck drivers and retail clerks. AI is good at learning how to do things people already do over and over again. But AI is not good at seeing new opportunities and inventing novel ways to do things. If you can always be inventive, the ghost of AI isn't likely to threaten you.

What, then, will work look like in coming years? Life is likely to be a constantly changing swizzle of jobs of various intensities, entrepreneurial ventures, freelance work, and side gigs. Instead of one career, you'll have many micro-careers. Instead of an eight-hour, five-day-a-week work schedule, you'll work all sorts of different hours on different days, from all sorts of different places. In a decade or two you might make extra money from selling power from your rooftop solar panels, or maybe you'll buy a self-driving car and let it work independently of you for Uber when you don't need it. If you add up all those activities, you'll be a constantly evolving, multifaceted, entrepreneurial business of one.

As discussed earlier, it's vital that policymakers help the working population make the leap from the fading world of scaled-up, full-time jobs to the emerging world of unscaled entrepreneurial work. If society can't do that and if AI catches up to one kind of routine work after another, we're going to be left with a vast unemployed class—at which point we'll need to consider some kind of universal basic income.

A big part of the solution is likely to be online education like that offered by Khan Academy, but it's a complex problem we must tackle. Just as schools and other institutions were developed generations ago to guide a once-agrarian population into Industrial Age life, new entities are going to need to be developed to guide the population into life in the entrepreneurial age.

* * *

This brings us back to education. If your career is turning into a life-long entrepreneurial business of one, how do you think about school and learning?

My kids are in a school I helped develop with Sal Khan—it's actually in the same Mountain View, California, building as Khan Academy's headquarters. We're trying to shape the way a classroom should work to prepare students for the next era. It aggressively adopts technology to help students learn academic material at their own pace, much as I described in the chapter on education. The school then encourages students to work on problems in teams so they learn how to collaborate—as they would if starting a business or solving a tough problem for a project. The teachers become coaches and guides, showing the students *how* to learn and *how* to collaborate.

I believe students will still need to physically go to a classroom for the foreseeable future (versus remote learning from home through online classes), but the role of the classroom needs to change. I want my kids to learn social, emotional, and leadership skills from the teacher and other students. They then need to learn how to seek out knowledge from online sources from the likes of Khan Academy, classes from MIT or Stanford, or raw data and software. Such technology allows students to learn at their own pace and pursue topics they're passionate about.

Most of all, the classroom should pull all these elements together so students learn systems thinking. Math, geography, and history should not be silos of learning; students need to learn how to put all their learning together to solve real-world problems in collaboration with others. That's what systems thinking is all about.

Now, most people don't have the privilege of sending their kids to an experimental school like ours. Still, it's worthwhile to think about an approach to education that might help prepare our children for the next few decades. Every parent should push their children's schools to teach in a way that anticipates an unscaled economy and get away from the kind of teaching that has long prepared students for scaled-up office and factory work.

After high school, learning and work will need to become inter-twined. We will modify the idea that we learn when young and then just work until we're old. Chapter 5 was about lifelong education in the cloud, which will play a significant role in our careers. We will need to continuously keep up with changes in technology and new informa-tion. And we will choose to continuously learn to satisfy our shifting passions. Early in life you might love one kind of work but later want to learn a new field. The emerging online learning sector will allow us to dip in and out of serious education. It will be our career fuel, con-stantly feeding our working lives and making us better.

This is a better way to live. The scaled economy made us conform to it. We took jobs that served the scaled-up companies we worked for. A lot of people in those jobs have functioned on autopilot for years, doing what they had to do to get a paycheck. The unscaled economy, however, conforms to us. It gives us the opportunity to choose a pas-sion—develop a superpower—and then find a market for it so more of us will get to do the work we really want to do and learn the things we really want to learn.

* * *

Two technologies will have an outsized impact on work: artificial intelligence and virtual reality. How should you think about those developments?

Successful people in the AI century will focus on work that takes advantage of unique human strengths, like social interaction, creative thinking, decision making with complex inputs, empathy, and ques-tioning. AI cannot think about data it doesn't have; it predicts what you want to see on Facebook based on what you've already liked. It can't predict that you might like something that's entirely out of left field—something that seems incongruous with your past patterns. Only humans can think that creatively. As author Kevin Kelly says, the most valuable people in an age of push-button answers will be those who ask the most interesting questions.

AI's proponents say it will collaborate with us, not compete against us. Cancer researcher M. Soledad Cepeda has given talks about AI in her work. She says that AI software can analyze in two seconds the amount of data and text that a research assistant would need two weeks to plow through. That frees up the assistants to do more thoughtful work and speeds up the scientists' search for cures.

So I believe that the most successful, influential people of the next twenty years will be those who understand how to partner with AI. The technology can seem scary, but it is the most powerful tool humanity has ever built. The people who can hitch AI to their uniquely human thinking will solve great problems and give us creations we can barely fathom today. This my chief career advice to anyone: learn how to use AI to help you accomplish your dreams.

Oddly enough, although stories in the media raise alarms about AI, I'm more worried about what VR—virtual reality—might do to us.

There's little doubt that VR experiences will get good enough to rival real-world experiences. Just look at the movie-like realism of today's video games and imagine being immersed in such a visually rich and responsive world. Now imagine that other people are in that world with you and that you can talk with them, touch them, work with them, or play sports with them. VR is going to get that good in the 2020s.

Inside that world will be commerce. You'll be able to buy software-created versions of things you'll need in your virtual world—maybe clothes, faster reflexes, or entrance to a cool nightclub. Because there's commerce, there will be work. A version of you in a virtual world—your avatar—might sell those clothes, coach other avatars on how to use their software-enhanced reflexes, or run the cool nightclub.

It will be possible—tempting for a lot of people—to spend more and more time in those VR worlds. Think of how parents become concerned when a teen spends hours playing video games—the allure of VR worlds will be even greater, especially for people looking for an escape from the real world. If someone has a job and friends in a VR world and enjoys being there, why would they spend much time in the physical world? As science fiction such as Neal Stephenson's *Snow Crash* long ago predicted, we could start losing chunks of the population to VR.

That would raise some existential questions. If we create an alternate society in VR, what does that mean for physical society—even for basic human values? (Can you kill someone's avatar in a VR world without consequence? You can in a video game, but if the avatar is key to you, your job, your relationships—is that different?) What happens if it becomes too easy to retreat from the problems of the real world? Will we care as much about global warming, preserving beautiful landscapes, or even whether our neighbor could use a kind word? Will people find a replacement for human touch in VR, and will they then give up on that kind of physical connection to others?

VR is an amazing and promising technology. I'm sure I'll continue to invest in it and encourage people to understand it, to build companies on it and in it. VR will play an important role in education. I hope my grandchildren can learn about the Great Depression by virtually inhabiting it and feeling what it's like to try to get work under those conditions. It will be a fascinating new medium—for entertainment, sports, news. Go forth and be a part of the opportunities VR will create. Yet part of me also thinks: please don't make it *too* real.

* * *

I don't ask my children what they want to be when they grow up. Right now it's a futile question. The technologies we're unleashing and the forces of unscaling are changing the world more profoundly than any time since the early 1900s. Some of the work that is rewarding and lucrative today will be wiped out in twenty years. Entirely new kinds of work that we can barely imagine now will be the hot jobs of 2030. The best strategy anyone can have is to stay curious, ambitious, and adaptable. Plan to learn throughout life, to work in many kinds of jobs and have several different careers. Promise yourself that you'll pay attention to your passions and superpowers and follow them.

I don't even know what I want to be when I grow up. AI and unscaling will change my business of creating businesses. But I find that exhilarating. I envy the generation who got to be alive when the automobile went from toy to life-altering machine, when flying rushed from scientific experiment to something anyone could do, when learning about the world shifted from words on paper to sounds over the radio to moving images on the television. Now I get to be alive and part of a similar technological, commercial, and societal epoch.

Anyway, I'm an optimist. Economies of scale did us a great deal of good for more than a century. I believe the AI century's economies of unscale will be even better. The goal for these next twenty years is to build systems that allow us to be happy, do the things we want to do, create what we thought was impossible, and make the world peaceful and prosperous. Watching the news some days, that can seem far-fetched. But I believe we're at the beginning of a remarkable time, and I can hardly wait to be a part of it.

Acknowledgments

My work has always been a collaborative process, and this book was no different. I have many people to thank for making this happen.

First, I want to thank my wife, Jessica, for letting me carve out time away from our kids—Ajay, Arya, and Isabella—to do this on top of an already busy schedule, especially on weekends!

I am grateful for the support and input of my partners at General Catalyst. I'd also like to thank the entrepreneurs we've funded for their insights and their willingness to tell their stories, many of which you'll find in these pages.

For this book I want to particularly thank Claire Baker, Ronda Scott, Spencer Lazar, and Armaan Ali—all members of the General Catalyst team—for their input and help with everything from research to logistics.

Thanks, also, to Professor Jim Cash at Harvard Business School and Jim Hock, formerly with the US Department of Commerce, for helping with our thinking; to our agent, Jim Levine, for his guidance from early on; and to this book's editor, John Mahaney, for seeing the project through.

And thanks to Danny Crichton, who helped me develop the original paper on economies of unscale that I published at HBR.org in 2013.

And of course this book would not have been possible without my collaborator, Kevin Maney. It was Kevin who encouraged me to write this book after reading the aforementioned paper, and he played a critical role in pushing the thinking behind the book.

Hemant Taneja
Palo Alto, California, 2018

Notes

Chapter 1: The Great Unscaling

6 **"made it for people to stay healthy":** Glen Tullman, interview with Kevin Maney, November 3, 2016.

13 **more than 2.5 billion people in 2016:** Mary Meeker, "2016 Internet Trends Report," Internet Trends 2016—Code Conference, June 1, 2016, www.kpcb.com/blog/2016-internet-trends-report.

17 **eighteen months for the same price:** Actually, when Gordon Moore first described what ultimately became Moore's Law, he believed computing power would double every two years. The pace of change proved so rapid, however, that he revised it to every eighteen months.

19 **"resistance of the established paradigm":** Carlota Perez, *Technological Revolutions and Financial Capital: The Dynamics of Bubbles and Golden Ages* (Northampton, MA: Edward Elgar Publishing, 2003), 151.

22 **work currently done by humans:** Carl Benedikt Frey and Michael A. Osborne, "The Future of Employment: How Susceptible Are Jobs to Computerisation?" (Oxford: Oxford Martin School, University of Oxford, 2013), www.oxfordmartin.ox.ac.uk/downloads/academic /The_Future_of_Employment.pdf.

Chapter 2: The AI Century

27 **more than 4 million in 2015:** Solomon Fabricant, "The Rising Trend of Government Employment," National Bureau of Economic Research, New York, 1949.

28 **"never would have found on our own":** Ben Popper, "First Interview: Chris Dixon Talks eBay's Purchase of Hunch," *Observer*, November 21, 2011, http://observer.com/2011/11/chris-dixon -ebay-hunch.

29 **by 2020 and 80 billion by 2025:** Michael Kanellos, "152,000 Smart Devices Every Minute In 2025: IDC Outlines the Future of Smart

Things," *Forbes*, March 3, 2016, www.forbes.com/sites/michaelkanellos /2016/03/03/152000-smart-devices-every-minute-in-2025-idc -outlines-the-future-of-smart-things/#acfb0a34b63e.

30 **an average of fifty-two per house:** Michael Kanellos, "The Global Lighting Market by the Numbers, Courtesy of Philips," Seeking Alpha, October 23, 2008, https://seekingalpha.com/article/101408 -the-global-lighting-market-by-the-numbers-courtesy-of-philips.

32 **and just a trickle before that:** John Stackhouse, "Back Off, Robot: Why the Machine Age May Not Lead to Mass Unemployment (Radiologists, Excepting)," Medium, October 29, 2016, https:// medium.com/@StackhouseJohn/back-off-robot-why-the-machine -age-may-not-lead-to-mass-unemployment-radiologists-excepting -6b6d01e19822.

32 **Neural Dynamics and Computation Lab:** Surya Ganguli, interview with Kevin Maney, July 13, 2016.

33 **will revolutionize the legal profession:** Julie Sobowale, "How Artificial Intelligence Is Transforming the Legal Profession," *ABA Journal*, April 1, 2016, www.abajournal.com/magazine/article /how_artificial_intelligence_is_transforming_the_legal_profession.

33 **mapping and understanding the brain:** Courtney Humphries, "Brain Mapping," *MIT Technology Review*, www.technologyreview .com/s/526501/brain-mapping.

33 **more capable in many ways than humans:** Jeff Hawkins, interview with Kevin Maney, June 22, 2016.

35 **"I got hooked," Thompson says:** David Kosslyn and Ian Thompson, interview with Kevin Maney, July 1, 2016.

36 **another VR startup, Penrose Studios:** Kevin Maney, "Afraid of Crowds? Virtual Reality May Let You Join Without Leaving Home," *Newsweek*, July 30, 2016, www.newsweek.com/afraid-crowds-virtual -reality-without-leaving-home-485621.

36 **like we do today on the web:** Philip Rosedale, interview with Kevin Maney, August 19, 2016.

37 **47,856 miles of federal highway:** "Interstate Highway System," Wikipedia, https://en.wikipedia.org/wiki/Interstate_Highway_System.

39 **"manufacturing very quickly," Friefeld says:** Max Friefeld, interview with Kevin Maney, 2017.

40 **jobs, geopolitics, and the climate:** Bernard Mayerson, "Emerging
    Tech 2015: Distributed Manufacturing," World Economic Forum,
    March 4, 2015, www.weforum.org/agenda/2015/03/emerging-tech
    -2015-distributed-manufacturing.

45 *The Future Is Better Than You Think:* Peter H. Diamandis and
    Steven Kotler, *Abundance: The Future Is Better Than You Think* (New
    York: Simon and Schuster, 2015), Kindle location 268.

45 **"creative destruction will take place":** Perez, *Technological
    Revolutions and Financial Capital*, 153–154.

CHAPTER 3: ENERGY: YOUR HOME WILL HAVE ITS OWN CLEAN POWER PLANT

50 **"to the electric power system," he says:** Naimish Patel, interview
    with Kevin Maney, November 15, 2016.

52 **wood, corn stalks, and dried manure:** Max Roser, "Energy
    Production and Changing Energy Sources," Our World in Data,
    https://ourworldindata.org/energy-production-and-changing-energy
    -sources.

55 **his "Master Plan, Part Deux":** Elon Musk, "Master Plan, Part Deux,"
    Tesla, July 20, 2016, www.tesla.com/blog/master-plan-part-deux.

56 **"global citizen," Hietanen told reporters:** Tom Turula, "'Netflix of
    Transportation' Is a Trillion-Dollar Market by 2030—And This
    Toyota-Backed Finnish Startup Is in Pole Position to Seize It," *Business
    Insider*, July 2, 2017, http://nordic.businessinsider.com/this-finnish
    -startup-aims-to-seize-a-trillion-dollar-market-with-netflix-of
    -transportation--and-toyota-just-bought-into-it-with-10-million
    -2017-7.

58 **systems that others can connect to:** "Grid Modernization and the
    Smart Grid," Office of Electricity Delivery and Energy Reliability,
    https://energy.gov/oe/services/technology-development/smart-grid.

58 **in line with those presented here:** "MIT Energy Initiative
    Report Provides Guidance for Evolving Electric Power Sector,"
    Massachusetts Institute of Technology, December 15, 2016, https://
    energy.mit.edu/news/mit-energy-initiative-report-provides-guidance
    -evolving-electric-power-sector.

59 **1980s while efficiency has rocketed:** Ramez Naam, "Solar Power Prices Dropping Faster Than Ever," Ramez Naam, November 14, 2003, http://rameznaam.com/2013/11/14/solar-power-is-dropping -faster-than-i-projected.

59 **needed for the entire United States:** Rob Wile, "How Much Land Is Needed to Power the U.S. with Solar? Not that Much," *Fusion*, May 10, 2015, http://fusion.net/how-much-land-is-needed-to-power-the -u-s-with-solar-n-1793847493.

59 **never need to burn another atom of carbon:** Diamandis and Kotler, *Abundance*, Kindle 204. "Since humanity currently consumes about 16 terawatts annually (going by 2008 numbers), there's over five thousand times more solar energy falling on the planet's surface than we use in a year."

59 **invested in renewable energy technology:** Jeffrey Michel, "Germany Sets a New Solar Storage Record," *Energy Post*, July 18, 2016, http:// energypost.eu/germany-sets-new-solar-storage-record.

60 **will run entirely on renewable energy:** Quentin Hardy, "Google Says It Will Run Entirely on Renewable Energy in 2017," *New York Times*, December 6, 2016, www.nytimes.com/2016/12/06 /technology/google-says-it-will-run-entirely-on-renewable-energy -in-2017.html?_r=0.

60 **Bloomberg New Energy Finance:** Tom Randall, "World Energy Hits a Turning Point: Solar that's Cheaper Than Wind," *Bloomberg Technology*, December 15, 2016, www.bloomberg.com/news/articles /2016–12–15/world-energy-hits-a-turning-point-solar-that-s-cheaper -than-wind.

62 **who collaborates with General Motors:** Christopher Mims, "Self-Driving Hype Doesn't Reflect Reality," *Wall Street Journal*, September 25, 2016, www.wsj.com/articles/self-driving-hype-doesnt -reflect-reality-1474821801.

63 **Institute of Electrical and Electronic Engineers:** Kevin Maney, "How a 94-Year-Old Genius May Save the Planet," *Newsweek*, March 11, 2017, www.newsweek.com/how-94-year-old-genius-save -planet-john-goodenough-566476.

68 **in spending on the grid by 2025:** Katherine Tweed, "Utilities Are Making Progress on Rebuilding the Grid. But More Work Needs to

Be Done, *Green Tech Media*, May 11, 2016, www.greentechmedia
.com/articles/read/Utilities-Are-Making-Progress-on-Rebuilding
-the-Grid.

69   **CEO Pasquale Romano told *Forbes*:** Joanne Muller, "ChargePoint's
New Stations Promise Fast Charge in Minutes for Your Electric Car,"
Forbes, January 5, 2017, www.forbes.com/sites/joannmuller/2017
/01/05/chargepoints-new-stations-promise-fast-charge-in-minutes
-for-your-electric-car/#7a769dee492d.

CHAPTER 4: HEALTHCARE: GENOMICS AND AI WILL ENABLE PROLONGED HEALTH

72   **"bugs in him," Laraki jokes now:** Othman Laraki, interview with
Kevin Maney, December 30, 2016.

75   **Disease Control and Prevention (CDC):** "Life Expectancy," Centers
for Disease Control and Prevention, FastStats, www.cdc.gov/nchs/
fastats/life-expectancy.htm.

76   **"attributed to waste, fraud and abuse":** Richard D. Lamm and
Vince Markovchick, "U.S. Is on Fast Track to Health Care Train
Wreck," *Denver Post*, December 17, 2016, www.denverpost
.com/2016/12/17/u-s-is-on-fast-track-to-health-care-train-wreck.

76   **"added financial pressure on consumers":** Gregory Curfman,
"Everywhere, Hospitals Are Merging—But Why Should You Care?"
Harvard Health Blog, April 1, 2015, www.health.harvard.edu
/blog/everywhere-hospitals-are-merging-but-why-should-you-care
-201504017844.

78   **"to go to Mecca—the big hospital":** Tullman, interview.

79   **"the safest or best place to be":** Paula Span, "Going to the
Emergency Room Without Leaving the Living Room," *New York
Times*, November 4, 2016, www.nytimes.com/2016/11/08/health
/older-patients-community-paramedics.html.

81   **president of the American Medical Association:** Laurie Vazquez,
"How Genomics Is Dramatically Changing the Future of Medicine,"
The Week, August 2, 2016, http://theweek.com/articles/639296/how
-genomics-dramatically-changing-future-medicine.

81   **"cancer immunology and immunotherapy":** Ibid.

82 **"or tying into lifestyle tools like Fitbit":** Jonathan Groberg, Harris Iqbal, and Edmund Tu, "Life Science Tools/Services, Dx, and Genomics," UBS Securities, May 2016.

84 **Cleveland Clinic's associate chief information officer:** Juliet Van Wagonen, "How Cleveland Clinic Stays on the Bleeding Edge of Health IT," *HealthTech Magazine*, March 9, 2017, https://healthtech magazine.net/article/2017/03/how-cleveland-clinic-stays-bleeding -edge-health-it.

84 **"a limited set of inputs and complexity":** Laraki, interview.

85 **J&J reluctantly stopped making them:** Anthony Cuthbertson, "Plug Pulled on Robot Doctor After Humans Complain," *Newsweek*, March 30, 2016, www.newsweek.com/plug-pulled-robot-doctor-after -humans-complain-442036.

85 **take them, according to *Nature*:** Nicholas J. Schork, "Personalized Medicine: Time for One-Person Trials," *Nature*, April 29, 2015, www. nature.com/news/personalized-medicine-time-for-one-person -trials-1.17411.

86 **"N=1 provides that laser-focused result":** Kevin Kelly, *The Inevitable: Understanding the 12 Technological Forces That Will Shape Our Future* (New York: Penguin Group, 2016), Kindle edition, location 3521.

88 **connected medical device startups:** "Redefining the Future of Medicine: 72 Medical Device Startups Advancing Treatment and Prevention," CB Insights, September 15, 2016, www.cbinsights.com /blog/brain-scans-pacemakers-72-medical-device-startups-market -map-2016.

89 **medical-record app using their own information:** "PatientBank Is Creating a Unified Medical Record System," Y Combinator, August 10, 2016, https://blog.ycombinator.com/patientbank.

90 **"diagnosis by the time they are an adult":** Kelly, *Inevitable*, Kindle 470.

90 **"to do something great for humanity":** Taylor Kubota, "Deep Learning Algorithm Does as Well as Dermatologists in Identifying Skin Cancer," *Stanford News*, January 25, 2017, http://news.stanford .edu/2017/01/25/artificial-intelligence-used-identify-skin-cancer.

91 **"'We are waiting for you'"**: "11 Health System CEOs on the Single
Healthcare Problem They Want Fixed Tonight," *Becker's Hospital
Review*, November 11, 2016, www.beckershospitalreview.com
/hospital-management-administration/11-health-system-ceos-on
-the-single-healthcare-problem-they-want-fixed-tonight.html.

92 **University of Pennsylvania, told *Nature*:** David Cyranoski,
"CRISPR Gene-Editing Tested in a Person for the First Time,"
Nature, November 15, 2016, www.nature.com/news/crispr-gene
-editing-tested-in-a-person-for-the-first-time-1.20988.

CHAPTER 5: EDUCATION: LIFELONG LEARNING FOR DYNAMIC AND PASSIONATE WORK

93 **"I was twelve to when I was eighteen":** Sam Chaudhary, interview
with Kevin Maney, January 13, 2017.

95 **"log in, and see how we are doing":** Ethan Forman, "ClassDojo App
Helps Danvers School Keep Things Positive," *Salem News*, March 27,
2017, www.salemnews.com/news/local_news/class
dojo-app-helps-danvers-school-keep-things-positive/article
_6cc9e48e-7496–56da-95e2-bb22b7992311.html.

97 **"places of learning," Gray concludes:** Peter Gray, "A Brief History
of Education," *Psychology Today*, August 20, 2008, www.psychology
today.com/blog/freedom-learn/200808/brief-history-education.

98 **has nearly fifty-nine thousand:** "List of United States University
Campuses by Enrollment," Wikipedia, accessed April 2017, https://
en.wikipedia.org/wiki/List_of_United_States_university_campuses
_by_enrollment.

98 **decline in new business startups:** Arnobio Morelix, "3 Ways Student
Debt Can Affect Millennial Entrepreneurs," Kauffman Foundation,
May 27, 2015, www.kauffman.org/blogs/growthology
/2015/05/3-ways-student-debt-can-affect-millennial-entrepreneurs.

100 **"go to college, become a professional":** Sal Khan, interview with
Kevin Maney, June 24, 2016.

103 **started working on Khan Academy courses:** "Research on the Use
of Khan Academy in Schools," SRI International, www.sri.com/work
/projects/research-use-khan-academy-schools.

107   **across much of Asia and Africa:** Harshith Maliya, "Can MOOC
      Platforms Galvanise Universal Education in India?" *Your Story*, April
      28, 2017, https://yourstory.com/2017/04/coursera-nikhil-sinha.

CHAPTER 6: FINANCE: DIGITAL MONEY
AND FINANCIAL HEALTH FOR ALL

109   **"what pain we get from it as well":** Ethan Bloch, interview with
      Kevin Maney, July 11, 2016.

110   **they were spending down their assets:** Eugene Kim, "A 29-Year-Old
      Invented a Painless Way to Save Money, and Google's Buying into It,"
      Business Insider, February 19, 2015, www.businessinsider
      .com/digit-ceo-ethan-bloch-interview-2015-2.

112   **"could handle 1,200–1,500 checks per hour":** "A History of Federal
      Reserve Bank of Atlanta, 1914–1989," Federal Reserve Bank of
      Atlanta, www.frbatlanta.org/about/publications/atlanta-fed-history
      /first-75-years/the-bank-in-the-1960s.aspx.

112   **offered the first universal credit card:** "Credit Card," Encyclopedia
      Britannica, www.britannica.com/topic/credit-card.

113   **less than 10 percent of all US assets:** Steve Schaefer, "Five Biggest
      U.S. Banks Control Nearly Half Industry's $15 Trillion in Assets,"
      Forbes, December 3, 2014, www.forbes.com/sites/steveschaefer/2014
      /12/03/five-biggest-banks-trillion-jpmorgan-citi-bankamerica/#6db
      9672db539.

120   **and the JOBS Act is an example:** David Pricco, "SEC's New Jobs
      Act Title III Crowdfunding Rules: Overview and First Thoughts,"
      Crowdexpert, http://crowdexpert.com/articles/new_jobs_act_titleiii
      _rules_overview_first_thoughts.

121   **Paris Hilton bought into ICOs:** Mike Orcutt, "What the Hell Is an
      Initial Coin Offering," *Technology Review*, September 6, 2017,
      technologyreview.com/s/608799/what-the-hell-is-an-initial-coin
      -offering.

121   **China outright banned the offerings:** Jon Russell, "First China,
      Now South Korea Has Banned ICOs," *Techcrunch*, September 28,
      2017, https://techcrunch.com/2017/09/28/south-korea-has-banned
      -icos.

CHAPTER 7: MEDIA:
CONTENT YOU LOVE WILL FIND YOU

125   **together a vast array of entertainment:** Trey Williams, "More People
      Subscribe to a Streaming Service than They Do Cable TV,"
      MarketWatch, June 9, 2017, www.marketwatch.com/story/more
      -people-subscribe-to-a-streaming-service-than-they-do-cable-tv-2017
      -06-09.

129   **"the anchor for live audio":** John Donham, interview with Kevin
      Maney, June 22, 2016.

129   **head of strategy for Katz Media Group:** Stacey Lynn Schulman,
      "A Closer Look at the Future of Radio," *Radio Ink*, June 30, 2016,
      http://radioink.com/2016/06/30/closer-look-future-radio.

129   **peaked in 1909 at twenty-six hundred:** "Newspapers,"
      Encyclopedia.com, www.encyclopedia.com/literature-and-arts
      /journalism-and-publishing/journalism-and-publishing/newspaper.

130   **scale in media was in full swing:** Sam Lebovic, "The Backstory of
      Gannett's Bid to Buy Tribune," *Columbia Journalism Review*, April 29,
      2016, www.cjr.org/business_of_news/frank_gannett_robert
      _mccormick_and_a_takeover_bids_backstory.php.

131   **in American homes and businesses:** "Television Facts and
      Statistics—1939 to 2000," Television History—The First 75 Years,
      www.tvhistory.tv/facts-stats.htm.

132   **apex of scaled-up media giants:** Lara O'Reilly, "The 30 Biggest
      Media Companies in the World," *Business Insider*, May 31, 2017,
      www.businessinsider.com/the-30-biggest-media-owners-in-the-world
      -2016-5/#28-prosiebensat1-291-billion-in-media-revenue-3.

132   **rounds of buyouts and layoffs:** "State of the News Media 2016,"
      Pew Research Center, June 15, 2016, www.journalism.org/2016/06
      /15/state-of-the-news-media-2016.

133   **earns $7.5 million annually:** Annette Konstantinides, "Nice Work if
      You Can Get It: The World's Highest-Earning YouTube Stars Who
      Make Up to $15m a Year from Their Online Shows," *Daily Mail*,
      December 6, 2016, www.dailymail.co.uk/news/article-4007938/The
      -10-Highest-Paid-YouTube-stars.html.

133   **2012 to 33 percent in 2015:** "More Americans Using Smartphones for Getting Directions, Streaming TV," January 29, 2016, Pew Research Center, www.pewresearch.org/fact-tank/2016/01/29 /us-smartphone-use.

136   **public trusted the mainstream media:** "Americans' Trust in Mass Media Sinks to New Low," Gallup, September 14, 2016, www.gallup .com/poll/195542/americans-trust-mass-media-sinks-new-low.aspx.

138   **clues that were all AR overlays:** Raymond Winters, "Augmented Reality: Commercial and Entertainment Applications," Nu Media Innovations, June 29, 2016, www.numediainnovations.com/blog /augmented-reality-commercial-and-entertainment-applications.

CHAPTER 8: CONSUMER PRODUCTS: EVERYTHING YOU BUY WILL BE EXACTLY WHAT YOU WANT

140   **"we learned a little bit about Luxottica":** Max Chafkin, "Warby Parker Sees the Future of Retail," *Fast Company*, February 17, 2015, www.fastcompany.com/3041334/warby-parker-sees-the-future -of-retail.

142   *the Age of (Nearly) Perfect Information:* Itamar Simonson and Emanuel Rosen, *Absolute Value: What Really Influences Customers in the Age of (Nearly) Perfect Information* (New York: HarperBusiness, 2014).

143   **about 60 percent of global GDP:** "List of Largest Consumer Markets," Wikipedia, https://en.wikipedia.org/wiki/List_of_largest _consumer_markets.

146   **commercial real-estate firm CoStar:** Ashley Lutz, "The American Suburbs as We Know Them Are Dying," *Business Insider*, March 5, 2017, www.businessinsider.com/death-of-suburbia-series-overview -2017-3?IR=T.

146   **share of second-place Nordstrom:** Jason del Ray, "Millennials Buy More Clothes on Amazon than Any Other Website," *Recode*, March 9, 2017, www.recode.net/2017/3/9/14872122/amazon-millennials -online-clothing-sales-stitch-fix.

147   **compiled by Wessels Living History Farm:** "Shrinking Farm Numbers," Wessels Living History Farm, www.livinghistoryfarm .org/farminginthe50s/life_11.html.

CHAPTER 9: POLICY:
PROFOUND DECISIONS TO HAVE A GOOD AI CENTURY

159 **said when the group was formed:** David Gershgorn, "Facebook, Google, Amazon, IBM, and Microsoft Created a Partnership to Make AI Seem Less Terrifying," *Quartz*, September 28, 2016, https://qz.com /795034/facebook-google-amazon-ibm-and-microsoft-created-a -partnership-to-make-ai-seem-less-terrifying.

159 **with machines doing the rest:** Nanette Byrnes, "As Goldman Embraces Automation, Even the Masters of the Universe Are Threatened," *MIT Technology Review*, February 7, 2017, www .technologyreview.com/s/603431/as-goldman-embraces-automation -even-the-masters-of-the-universe-are-threatened/?utm_campaign =add_this&utm_source=twitter&utm_medium=post.

159 **CEO of Kensho, told the *New York Times*:** Nathaniel Popper, "The Robots Are Coming for Wall Street," *New York Times*, February 25, 2016, www.nytimes.com/2016/02/28/magazine/the-robots-are -coming-for-wall-street.html?_r=1.

160 **the Bureau of Labor Statistics:** "Occupational Changes During the 20th Century," Bureau of Labor Statistics, www.bls.gov/mlr/2006 /03/art3full.pdf.

164 **"critical sectors," the group concluded:** Yale Law School, Information Society Project, https://law.yale.edu/isp.

165 **lawmakers from imposing government oversight:** "ANSI: Historical Overview," ANSI, www.ansi.org/about_ansi/introduction/history.

166 **Alzheimer's, Parkinson's, and celiac:** Julia Belluz, "In an Amazing Turnaround, 23andMe Wins FDA Approval for Its Genetic Tests," *Vox*, April 6, 2017, www.vox.com/2017/4/6/15207604/23andme -wins-fda-approval-for-its-genetic-tests.

166 **social issues raised by this research:** "Ethical, Legal and Social Issues in Genomic Medicine," National Human Genome Research Institute, www.genome.gov/10001740/ethical-legal-and-social -issues-in-genomic-medicine.

167 **patients to choose to try those drugs:** Adam Thierer and Michael Wilt, "The Need for FDA Reform: Four Models," Mercatus Center, September 14, 2016, www.mercatus.org/publications/need-fda -reform-four-models.

167   **wouldn't recommend it to their peers:** Erin Dietsche, "10 Things to Know About Epic," *Becker's Hospital Review*, January 20, 2017, www .beckershospitalreview.com/healthcare-information-technology /10-things-to-know-about-epic.html.

CHAPTER 10: THE CORPORATION: CHARTING AN UNSCALED FUTURE FOR SCALED ENTERPRISES

174   **swarm of bees taking down a bear:** CB Insights, "Disrupting Procter & Gamble: Private Companies Unbundling P&G and the Consumer Packaged Goods Industry," April 19, 2016, www.cbinsights.com /blog/disrupting-procter-gamble-cpg-startups.

175   **elements discovered through Connect + Develop:** Nesli Nazik Ozkan, "An Example of Open Innovation: P&G," *Science Direct*, July 3, 2015, www.sciencedirect.com/science/article/pii /S1877042815039294.

176   **"organization (or even to GE Predix!)":** "Improving Speed to Development—Lessons Learned While Building Aviation Apps on Predix," Predix Developer Network, January 12, 2017, www.predix .io/blog/article.html?article_id=2265.

177   **more than five hundred people:** "Frequently Asked Questions," Small Business Administration, www.sba.gov/sites/default/files /FAQ_Sept_2012.pdf.

177   **3.5 percent for the Amazon retail business:** Alexei Oreskovic, "Amazon Isn't Just Growing Revenue Anymore—It's Growing Profits," *Business Insider*, April 28, 2016, www.businessinsider.com /amazons-big-increase-in-aws-operating-margins-2016-4.

181   **"I spend time thinking about this topic":** Jeff Bezos, "2016 Letter to Shareholders," Amazon.com, April 12, 2017, www.amazon.com /p/feature/z6o9g6sysxur57t.

CHAPTER 11: THE INDIVIDUAL: LIVING YOUR LIFE AS A PERSONAL ENTERPRISE

188   **ability to work wherever they want:** "Multiple Generations @ Work," Future Workplace, http://futureworkplace.com/wp-content/uploads /MultipleGenAtWork_infographic.pdf.

188 **companies instead of big corporations:** Joshua Reeves, interview
     with Kevin Maney, April 15, 2017.
191 **ask the most interesting questions:** Kelly, *Inevitable*, Kindle 4211.
191 **given talks about AI in her work:** Darius Tahir, "IBM to Sell
     Watson's Brainpower to Speed Clinical and Academic Research,"
     Modern Healthcare, August 28, 2014, www.modernhealthcare.com
     /article/20140828/NEWS/308289945.

Index

Index

Also by Susan Lee

A New Economic View of American History
(with Peter Passell)

Susan Lee's

ABZs
of
ECONOMICS

Introduction by
Leonard Silk
Illustrations by Robert Mansfield
Poseidon Press • New York

For my mother

Published by Poseidon Press,
A Division of Simon & Schuster, Inc.
Simon & Schuster Building
Rockefeller Center
1230 Avenue of the Americas
New York, New York 10020

POSEIDON PRESS is a registered trademark of Simon & Schuster, Inc.

Designed by Barbara Marks
Manufactured in the United States of America

10 9 8 7 6 5 4 3 2 1

Library of Congress Cataloging-in-Publication Data

Lee, Susan, 1943–
  Susan Lee's ABZs of economics.

  1. Economics—Dictionaries. I. Title.
II. Title: ABZs of economics.
HB61.L34 1987   330'.03   86-25481
ISBN: 0-671-55711-4

INTRODUCTION

Susan Lee is one of a kind—a natural-born teacher, an irreverent and sassy comedienne (a menace to the pompous on the television talk shows and an eviscerator of party lines, whoever's party line it might be), and a skillful economist with a mind and style of her own. Best of all, for the sake of the present business in hand, she writes well—lucidly and simply and gutsily. This kind of writing stems not just from knowledge and technical skill but from a genuine desire to reach her readers and help them.

Lord knows, the public needs all the help it can get if it is to understand what economics is all about. This is a science that cannot be walled off from other sciences or from the broad social environment. Economists themselves sometimes forget that or seek to so limit the terrain of economics as to exclude all considerations except the ones they want to focus on. In the old days, one of the favorite characters of the economics teacher was Robinson Crusoe, alone on his desert island, trying to decide how much of his labor to devote to gathering coconuts, how much to building a canoe that would enable him to catch more fish in the long run. A slight increase in the complexity of the environment occurred when Friday showed up, and the relations between capital and labor could be explored.

Actually, figuring out how Crusoe could best serve his own

welfare, with or without Friday, was complicated enough for some of the best minds in the early days of economics; and the result of their desert-island analysis still has relevance to the economics of teeming cities, national and international economies. For the simple Crusoe model clarified the choices an economic agent (whether an individual, a business, or a nation) needs to consider before deciding how best to serve his, her, or its welfare. That process of weighing and deciding what is in one's own best interest is what we mean by rational behavior. In fact, is human behavior rational? Economics usually assumes that it is, or ought to be. It would be difficult, perhaps impossible, to invent an economics of abnormal and irrational behavior. But, alas, that is not to say that a good deal of human behavior, even in the economic area, is not irrational—that is, likely to frustrate and defeat one's own purposes, hurt one's own well-being, and injure the interests of others, even those one does not want to hurt.

Human behavior—whether rational or irrational—is subject to many constraints. Some are technical, such as the "law of diminishing returns"—the more fertilizer you put on a given amount of land, the less extra output of wheat or corn you get from each extra dose of inputs. Whether you're sane or insane, whether you're Reagan or Qaddafi, that holds.

Some constraints are legal: You might increase your income, create a monopoly, and maximize your economic well-being by putting out contracts to have your competitors eliminated by gunmen, but the law says this is a no-no, and you might fry for it. All the same, it is sometimes done.

Some constraints are moral: Thou shalt not sexually harass thy employees, even if there is no law saying you can't; morality patrols an even vaster terrain than law but is an even more uncertain and subjective patrolman; consciences differ from person to person, and hypocrisy often vitiates proclaimed principles of morality, even when the principles are commonly held and asserted. It was the great French cynic, the Duc de La Rochefoucauld who said, "Hypocrisy is the tribute that vice pays to virtue."

But the greatest constraint on human behavior, as the mathematician John von Neumann and the economist Oskar Morgenstern, the creators of game theory, demonstrated, is the realization that what you want to do will cause other people to do something back to you. The sergeant on "Hill Street Blues" used to send the station's policemen and -women out with the injunction, "Do it to them before they do it to you." But the wiser and more exact message would be, "Before you do it to them, consider what they may do to you—and then consider what we may have to do, given all the factors, legal, political, social, financial, technical, and so on, that affect our behavior." Anybody who thinks I am overstating the complexities of police behavior has not, I suspect, had recent experience in trying to get the cops to deal with a burglary or some other crime.

So economics is a slice of life, not the whole of it, but a big bloody slice. It plays a crucial role in the everyday lives of individuals, businesses, and nations, and should help you cope better with life's daily trials and confusions. That's where Susan Lee's dictionary comes in. I think it can even help you win some of the games.

—LEONARD SILK

PREFACE

Interest in things economic seems to be everybody's dirty little secret. And few know that better than an economist. For several years of my life, for example, I had to answer the question, "What do you do?" by saying, "I teach economics." Despite the fact that my reply was always apologetic, the response was inevitable: eyes rolled, faces darkened, heads shook, and lectures on why economics was deficient, silly, or boring were given. Nonetheless, after all that came the question: "As an economist, what do you think of (fill in appropriate hot topic)?" The interest was as inevitable as the disapproval. The perfect dirty little secret.

At the risk of revealing myself as a pornographer, then, allow me to purvey my goods.

1. Interest in economics is natural. Everybody is an economic agent in some sort of economic system. Whether it's a consumer shopping for the lowest price or a seller looking for the highest bid, economics forces its way into lots of everyday decisions. It's hard to resist being curious about something so intrusive.

2. Interest in economics is desirable. Most everybody wants to make their money earn money. But being a good investor depends, in part, on knowing something about the economic climate. It's good to be interested in something you can profit by.

3. Interest in economics is important. Being a citizen of the world requires an intelligent rendering of the news. Knowing some economics is essential to a grasp of what's going on, who is doing what to whom, and why.

If all this is so, then why so much tut-tutting about economics (and pooh-poohing of economists)? Most obviously, economics is not always easy to figure. It is filled with lingo and buzzwords. Too, economics has gotten some bad press. There's that sobriquet, "the dismal science," and the fact that economists are endlessly disputing one another. In sum, it's easy to dismiss economics as a mug's game. Easy, but not smart.

And that's where this book comes in. The *ABZs* is a collection of the important buzzwords in economics. It does several things: provides a straightforward definition of the terms, an explanation of the ideas behind the terms, and, where appropriate, a glimpse into how the ideas fit into public policy debates.

The *ABZs* isn't meant to be read straight through. Nonetheless, if a question sends you to the page where inflation is explained, you might also have to turn to the page where deflation is described to get a more complete answer.

The *ABZs* won't turn you into an economist (and that's probably the last thing you want). But it can help you glide over some of the rough patches in your data base, help you make more sense of the daily headlines, and increase your understanding of why economic questions seem to pop up so often.

Absolute Advantage

A term beloved by international trade types. A country has an absolute advantage when it can produce something more efficiently than another country. Mexico, for example, has an absolute advantage over Canada in growing tomatoes. The possession of various absolute advantages makes trade among countries attractive. Why should Canada spend its resources producing tomatoes when lower-cost tomatoes can be had from Mexico?

Absolute advantage can apply also to firms or to individuals. And it makes good sense here, too: If one person makes delicious bread but is unable to hammer a nail deftly, then it's reasonable for that person to specialize in making bread and to purchase the services of a carpenter who can hammer nails. See COMPARATIVE ADVANTAGE.

Aggregate Demand

Simply the sum of all the separate demands in an economy for goods and services.

Aggregate demand is usually divided into four components:

One, the demand of households for goods and services. This is straightforward enough—households buy all sorts of things, from pork roasts to visits from the TV repairperson.

Two, the demand of firms and governments for investment goods. This can be anything from a hammer to a sophisticated computer.

Three, the demand of government for goods and services. Think of the government as a combo household-business: it needs a wide range of things, from flowers for the White House Rose Garden, secretarial services for Congresspeople, to complicated missiles for the Defense Department.

And four, the demand of foreign households, firms, and governments for goods and services. This can be anything from American designer jeans to American maritime insurance. See DEMAND.

Allocation of Resources

What every beginning textbook in economics says the science of economics is all about.

Resources are things like land, water, air, labor, machinery, and even technical know-how. These resources are said to be "scarce" in the sense that their availability is limited, although some resources are more limited than others. An economic system exists to distribute these resources, and an efficient economic system is one that allocates them best—that is, where resources are employed in producing the stuff that people want most and not in producing things that nobody wants.

At one extreme is a planned economy where state bureaucrats make the decisions to allocate resources. At the other extreme is a free market economy where prices determine the allocation. Most economies fall somewhere in between.

Antitrust

The "trust" part is a quaint way of describing monopolies, restraints of trade, or conspiracies to prevent competition. It refers to the turn-of-the-century practice of merging small companies into one huge one in order to limit supply and raise prices. (Most American history texts brim with colorful accounts of this activity—from the Railroad Trust and the Oil Trust to the Tobacco Trust.) The "anti" part signaled the end of the government practice of granting exclusive rights to one firm to carry on its business.

Antitrust laws made their appearance in the United States in 1889: The first major antitrust law was the Sherman Anti-Trust Act of 1890, which was followed by the Clayton Anti-Trust Act in 1914, the Robinson-Patman Act in 1936, and the Celler-Kefauver Amendment in 1956.

These laws cover all sorts of business activities that are thought to involve unfair forms of competition. Among the taboos are companies getting together to decide on a common price for their product (price-fixing), companies selling

their products at varying prices in order to prevent competition in a certain market (price discrimination), companies pricing their products so low that potential competitors are discouraged from entering the market (predatory pricing), and certain types of mergers that might lead to one company dominating an industry. Firms found in violation of antitrust laws can find themselves compelled to pay rather stiff civil, and even criminal, penalties.

Some critics now contend that antitrust has gone too far. They have several reasonable arguments. First, monopolies aren't always bad—they often generate more rapid technological change and innovation than do firms that are busy competing with each other on a daily basis. Likewise, mergers that result in a firm dominating an industry often mean lower prices for consumers because the new, larger firm can produce more cheaply. Second, things aren't always as they seem —companies that appear to have a monopoly don't when the competition provided by foreign firms is included. Likewise, companies engaging in activities like price-cutting may be just aggressive competitors rather than potential monopolizers. And finally, critics can point to companies (like IBM) that have been prosecuted simply because they were "too big" rather than because they engaged in monopolistic activities, and to laws that were designed to protect small businesses but actually foster monopolistic situations.

Regardless of the economic or legal niceties, the fact remains that antitrust plaintiffs can make a pile of money if their lawsuits are successful. For example, a private firm that can prove damages from anticompetitive practices can sue for triple the losses actually suffered. So, too, the federal government has made a heavy-duty commitment to root out violators: the Antitrust Division of the Justice Department employs a slew of lawyers (and economists) to pursue alleged culprits with energy. And with fortitude—antitrust cases can take years before they even come to trial. See COMPETITION; MONOPOLY.

Appreciation

A rise in the price of a currency relative to other currencies. Under a system of floating exchange rates, the market sets the value of a currency—it's worth whatever buyers will pay for it. A nation with a "strong" currency—one that other nations wish to hold—will see its currency increase in value relative to other currencies. The nation with the appreciating currency usually then finds itself with a trade deficit because, as its exports become more expensive and its imports become less expensive, it imports more than it exports. See BALANCE OF PAYMENTS; FOREIGN EXCHANGE.

Arbitrage

A sophisticated version of buying cheap and selling dear. Arbitrage involves the act of buying something in one market and simultaneously selling it for more money in another market. Arbitraging is usually done by professional traders in the rarefied, high-tech world of foreign currency markets, or the spot market for commodities. But it can be accomplished anywhere and by anybody plucky—and lucky—enough to spot a favorable price discrepancy.

Opportunities for arbitrage are what bring prices in var-

ARBITRAGE

ious markets more or less in line with one another: when traders remove supply from cheap markets, they bring up prices there, and when traders deliver supply to expensive markets, they push down prices. Thus, arbitraging tends to equalize prices in all markets.

Asset

Can be either a physical or intangible property that has value to the owner. Individuals own assets in the form of houses or perfect credit records, firms own assets in the form of machines or copyrights. The asset's value lies in its use in providing future service or benefit: Homeowners can live in or sell their house; creditworthy individuals can borrow against their unblemished record.

The formal designation, "assets," is typically found on a firm's balance sheet. Here, assets are divided into "current" (either cash or things that can be readily turned into cash) and fixed (either buildings and equipment) or intangibles (copyrights, patents, and goodwill). Assets appear on one side of the balance sheet, liabilities on the other. See LIABILITY.

Austrian School

The roots of this school go deep into the nineteenth century with founder Karl Menger (1840–1921). Menger's chief contribution to economic thought lies in his exploration of the theory of value, particularly his notion that utility, or pleasure, declines with each extra unit of consumption. (Simple, really: The second consecutive banana split is not quite as good as the first.)

Today, however, the Austrian School is known by the work of two Menger disciples, Ludwig von Mises and Friedrich August von Hayek. Due to the work of these two men, the term "Austrian School" has become synonymous with laissez-faire capitalism.

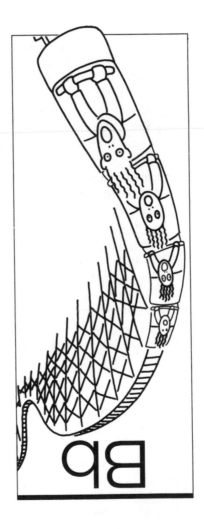

Balance of Payments

A phrase used to describe the amount of money flowing in and out of a country.

The phrase itself is associated with some rather colorful images: from well-tailored international bankers flying about the world lending money to sleazy speculators "parking" money in various countries overnight. The reality, however, is rather less dramatic. Balance of payments involves a complex set of activities, including foreign exchange operations, which are mostly humdrum and always exceedingly difficult to grasp at one sitting.

For starters, the flows of investment money described above are only a small part of the world's payment transactions. Another—and a larger part of the U.S. balance of payments—involves flows of goods: exports of nuclear turbines or imports of wristwatches. These transactions are logged under the category called the current account. In addition to merchandise flows, the current account keeps track of exports and imports of services—transportation, travel, and financial and insurance services included. It also encompasses big chunks of government spending on military business and foreign aid, smaller chunks for Social Security recipients living abroad, and dividend and interest payments.

A second category, called the capital account, logs three types of flows. The long-term account consists of investments such as Japanese automobile factories in Ohio, American bank loans to Brazilian businesses, and transactions for stocks and bonds. The short-term account measures the investments known as "hot money" that zip from country to country in search of quick profits. And the official reserve account consists of the purchases and sales of gold and foreign currencies by the Federal Reserve Bank, and the dollars held by central banks of other countries. Unlike other transactions, activity in the official reserve account is not related to trading goods or services, making transfer payments, or investing. Rather, activity here "adjusts" the exchange rates among different currencies. And that brings the discussion around to the role

of the foreign exchange markets in the balance of payments.

Simply put, foreign exchange markets are crucial to all this flowing about of goods and services because different countries have different currencies. If the Indian government wants to buy some nuclear turbines from a U.S. firm, it must first exchange its rupees for American dollars. Ditto for an American firm wishing to buy Japanese watches—dollars must be exchanged for yen. That seems very straightforward. And it will stay straightforward as long as countries import about as much as they export. But now consider what happens on the foreign exchange market when a country becomes a net importer; that is, when it buys more abroad than it sells, and thus runs a deficit in its balance of payments accounts.

Obviously, the importing country must exchange its money into foreign currencies in order to make its purchases. But because foreign countries are selling to the importing country rather than buying from it, they have little need for its currency. Since other countries don't want the importing country's currency, the value of its currency will decline, or depreciate, relative to other currencies. This effect holds in reverse, too: A country whose goods are in huge demand by other countries will find that its currency will appreciate compared to other currencies.

But the foreign exchange market is more than just an automatic adjustment mechanism among currencies. It is also made up of people who simply trade currencies, hoping to turn a profit by sudden moves in value. These traders watch a country's balance of payments as an early indicator of whether its currency will weaken or strengthen. Consider, for example, what happens when a country starts to run a deficit in its balance of payments. If traders rush to unload the hapless currency, its value will drop unless the central bank, using the resources in its official reserve account, steps in to buy its own currencies on the foreign exchange market. (This is called intervention in the exchange markets, and international types love to debate its usefulness in promoting something called stability.)

By now the intrepid might be wondering why all this back and forth, up and down, matters. The short answer is that it matters because countries are proud when their currencies are strong and embarrassed when they are weak. The long answer is that wide swings in a country's currency can upset trading relationships, causing grief to businesses in both the import and export sector. See FOREIGN EXCHANGE; GOLD STANDARD.

Balance Sheet

Document showing a firm's assets and liabilities at one particular point in time, usually the end of the year. It's called a balance sheet because assets equal liabilities; in fact, it's almost an arithmetical tautology.

Consider the balance sheet of a small firm, The Framistan Company: Assets, which consist of all the stuff Framistan owns, typically include the cash on hand, accounts receivable from framistans sold, inventories of both the materials to make framistans and the framistans made but not sold, and the factory and equipment needed to make framistans. Liabilities, which consist of all the stuff Framistan owes, typically include accounts payable to suppliers, salaries payable to workers, managers, and officers of the company, the mortgage on the factory, and, of course, any loans outstanding.

If The Framistan Company is a healthy enterprise, then its assets will add up to more than its liabilities. But that would unbalance the balance sheet. So the difference between assets and liabilities, called owners' equity (or net worth), is added to the liability side to make things equal. Why the liability side? Because owners' equity represents what is owned by the stockholders in The Framistan Company; thus, owners' equity is a liability—it is owed to the owners.

There is, however, one important element not reflected on Framistan's balance sheet—goodwill. This pleasant-sounding entry reflects Framistan's past success and future prospects: the high spirits of its employees, the utter respectability of its

name, the good relationship it enjoys with its suppliers and customers. How does one put a price on goodwill? Simple. A buyer wishing to purchase The Framistan Company will probably have to pay much more than its owners' equity— that sum is then called goodwill.

The Framistan Company is a very stripped-down example. The balance sheet of a larger company, or one with fancy accountants, contains many more categories: Assets might include marketable securities and other investments; liabilities might include dividends payable and accrued taxes. And owners' equity might include preferred, common, and treasury stock, along with retained earnings.

Bank Failure

The polite way of designating a bankrupt bank. Banks are heavily regulated institutions—both their borrowing and lending practices are subject to scrutiny by state and federal agencies. Nonetheless, there are dozens and dozens of ways for a bank to get into serious trouble, from making a bunch of bad loans (not enough money coming in) to paying too high interest rates in order to attract deposit money (too much money going out). And, of course, a bank can also fail if its officers or staff engage in criminal mischief—like embezzling.

A bank is said to fail when it becomes insolvent; that is, when it doesn't have enough assets to cover its liabilities. But a failure can be an elusive event. Bank regulators have several options: They can prop up a technically failed bank by pumping in fresh money, as the Federal Reserve did in the case of Continental Illinois in 1984, or by issuing pieces of paper that banks can carry on their balance sheets as assets; or—and this is the preferred option—they can "merge" the failed bank with a healthy bank.

Before the 1930s, when bank regulatory agencies were put in place, a bank failure was a serious event. Back then, a failed bank was a failed bank—it closed its doors, leaving its depositors out in the cold, sometimes literally. Back then, even the

rumor of a bank in trouble was enough to propel depositors there, and at other banks, into a rush to withdraw their funds. That, of course, only exacerbated the situation, so that healthy banks failed along with troubled ones. (This sequence of events was called a bank panic or a run on the banks.)

Nowadays, however, the existence of bank regulators and deposit insurance has made the handling of bank failures a fairly routine thing. These days, the only activity bank failures excite is editorials frothing and foaming about the irresponsibility (kind) or stupidity (not so kind) of bankers. See BANKING; FEDERAL RESERVE SYSTEM.

Bankruptcy

When an individual or corporation is legally declared insolvent by a court of law. A declaration of bankruptcy is more than just a statement, however. Insolvency doesn't (usually) describe a situation as mean as flat broke, but rather one in which liabilities are greater than assets. In other words, there are assets enough to pay back some, but not all, debts. And that's where bankruptcy proceedings come in.

Bankruptcy is actually a process by which the assets of the bankrupt are transferred to a court-appointed official. The official then liquidates the assets and makes sure that creditors with legal priority (the Internal Revenue Service, for example) get paid off before the rest of the creditors, and that the creditors are all treated fairly. Bankruptcy can be voluntary or involuntary. Individuals and corporations can declare themselves bankrupt, or creditors can get together and petition the court to declare a bankruptcy.

Banks and Banking

Commercial banks do many things: First and foremost, they provide a service called financial intermediation, and earn profits by doing so.

A description of modern banking can get a little tedious,

but it is not complicated. Consider, first, how banks provide financial intermediation in the economy. Despite its fancy term, this service involves nothing more than linking up people who want to save money with people who want to spend it. Savers place their money in banks, and banks, in turn, lend the money to spenders. Without financial intermediation, savings would rest idle under mattresses and borrowers would not be able to buy the goods and services that create jobs and make the economy go round. Quite clearly, financial intermediation is crucial for a healthy, sophisticated economy.

Banks provide intermediation by offering inducements for savers to put their money in bank accounts rather than under mattresses. That is, banks make it worth savers' while to become lenders.

For starters, banks provide a convenient and safe place to stash money. Customers deposit their money, or open an account and receive checks, which act as a substitute for cash. Sure, a bank account is not quite as handy as stuffing dough under a mattress, but for large purchases, using bank checks is safer and less awkward than cash. Banks also offer savers a rate of return (interest) on savings, making them a better deal than mattresses.

After luring savers into depositing their money, banks can transform savers into lenders by allowing borrowers to use the funds. Banks make borrowers pay for this privilege by charging interest on the money they lend. If all goes according to plan, banks pay savers less interest on deposits than they charge borrowers for loans. Banks use this "spread" to pay their expenses, and anything left over constitutes profit. In this respect, then, financial intermediation depends on banks being able to borrow cheap and lend dear.

Financial intermediation covers a lot of transactions. On the savers' side, it encompasses everything from modest, long-term passbook accounts to the immodest sums deposited overnight by large corporations. On the borrowers' side, it encompasses modest, short-term consumer loans to immodest, long-term loans to foreign governments.

But financial intermediation is not the only thing that banks do. They are also important players for the creation of money. While banks can't exactly print it—the way the federal government can—they do something almost as neat.

Consider, now, what happens when a saver makes a bank deposit. The amount of the deposit, itself, say $1000, appears on the bank's balance sheet as a liability; the bank owes the money to the depositor. However, the bank doesn't keep 100 percent cash against the deposit, but only a fraction of it: hence the bank reserves (say) 10 percent of the initial deposit in cash and lends out the rest. (Since loans are assets, the $1000 deposit liability will be balanced by $100 in cash and $900 in loan assets.) The bank makes the loan by opening an account from which the borrower may draw cash or write a check—it comes to the same thing. In other words, the bank has used the first depositor's account to *create* $900 of money in a second account.

But money creation doesn't stop there. Presumably, the borrower will withdraw money on the new account to pay for whatever it was he or she wished to borrow the money for. That money then usually goes into a third depositor's account. Again, the bank will reserve 10 percent of that deposit as $90 cash and lend the rest, $810, to another borrower. Presto, another $810 has been added to the supply of money. And so it goes, involving many depositors and their banks, until the created loan money becomes too small to measure. (In this example, the initial $1000 deposit can create $10,000 in new deposits—$9000 in the form of loans and $1000 as cash reserves.) The fractional reserve system, thus, allows banks to increase the money supply. See BANK FAILURE; FEDERAL RESERVE SYSTEM.

Barter

Obtaining goods or services without using money. Without using *money*? Sure. Money is a latecomer to the world of trad-

ing things. Primitive people satisfied their acquisition im-
pulses by exchanging things like feathers for things like skins
—just as kids trade baseball cards. Whatever was owned could
be used to trade for ownership of something else.

But as civilization moved on, barter became too complicated to move with it. In fact, barter did itself in by encour-
aging specialization in production: If I produce super
tomatoes and you produce great broccoli, then it makes sense
for each of us to specialize and then trade our veggies. But
specialization makes us less able to satisfy all our needs. If,
for example, a sudden craving for bananas strikes, both of us
will have to find a banana farmer to gratify our desire. But
what if the banana farmer hates broccoli and is allergic to
tomatoes? We then have to hunt around for a farmer who
not only wants to trade her produce for our broccoli or tomatoes, but grows something that the banana farmer wants.
(The remedy for ever more cumbersome barter was the invention of a medium of exchange—money.)

All these difficulties notwithstanding, barter has made a
comeback in these presumably nonprimitive times. Whether
it's an informal trade of my hand-knit sweater for your
plumbing services, or a more formal organization where potential traders advertise their goods or services, bartering can
be cheaper than using the money economy. No need to pay
for the services of middlemen, and no need to pay taxes.

Barter is also done by businesses and countries. Indeed,
barter is very popular with governments of planned economies. Barter allows these nations to trade the stuff they regularly overproduce for stuff they regularly need, and all
without having to use the hard currency which they regularly
have a shortage of. Thus, Yugoslavia's overproduction of
canned ham becomes the means for acquiring airplanes from
McDonnell Douglas. (Barter may be making a comeback, but
it still creates difficulties—McDonnell Douglas ended up
serving some of the canned ham in its own cafeterias.)

Bretton Woods

Shorthand for a conference involving the world's most important industrial nations held in Bretton Woods, New Hampshire, in 1941. International honchos breathe the words Bretton Woods with extreme reverence. And with good reason. The cooperation of forty-five nations to achieve the grand goal of promoting trade and investment was a spectacular achievement.

The conferees created two institutions—the World Bank and the International Monetary Fund—and set up an international payments system. Essentially, the Bretton Woods arrangement revolved around a foreign exchange system that

BRETTON WOODS

was somewhere between fixed and floating rates, described as fixed but adjustable. The fixed part meant that currencies were pegged at a par value; the adjustable part meant that the currencies were allowed to fluctuate around that par value—but if they moved too far from par, their respective governments were supposed to adjust them.

Trouble was, when currencies moved in their favor, governments didn't have much incentive to adjust. And when currencies moved against them, they were afraid to adjust. Instead, governments just waited for a crisis to sort things out. Needless to say, the Bretton Woods arrangement was filled with crisis. And, in the early 1970s, after a particularly severe set of disturbances on the foreign exchange market, Bretton Woods was abandoned in favor of a more flexible system—called "managed floating rates." See FOREIGN EXCHANGE; INTERNATIONAL MONETARY FUND; WORLD BANK.

Budget, Federal Government

Just like any well-run household, the federal government keeps a yearly account of how much money goes out and how much comes in. On the expense side, the government records its purchases of goods—like bridges and filet mignon for White House dinners—along with its purchases of services— like salaries for government bureaucrats. On the revenue side, it records the payments it receives—mostly from taxes.

At the beginning of the fiscal year, in October, the various branches of government are supposed to set forth their spending limits in accordance with how much revenue is expected. While the budget goes through many estimates and revisions, nobody really has a fix on the numbers until the final bills and receipts are totaled at the end of the year.

There is, however one important (and peculiar) characteristic that distinguishes the federal budget from a household budget—the treatment of investment goods. When a citizen makes a long-term investment, like buying a house, the purchase price is not logged on the budget as a current expense:

Mortgage payments are surely a current expense, but not the total bill. Likewise, businesses segregate their capital spending on buildings and machines from their current spending like interest payments on the money borrowed to finance those capital projects.

The federal government, however, treats its purchases of investment goods just as if they were current expenses. Simply put, government spending for bridges and dams is lumped into the budget along with spending for bureaucrats' salaries and filet mignon. Thus, in a period of high capital spending—like a defense buildup—the expenditures for capital products weigh heavily on the federal budget, making the deficit look larger than if it were more conventionally calculated.

Also unlike any well-run household, the government has been spending a lot more than it takes in: hence the debut of what is often pronounced as one word, *budget deficit*. Now, deficits aren't necessarily a bad thing. In fact, many economists encourage the government to turn to red ink whenever the economy slumps. The idea is that if the government steps up its buying of goods and services, then unemployment won't be as severe as it otherwise might be.

The corollary to a budget deficit is, of course, a *budget surplus*. Here, the notion is that during good economic times the government should pull back on its own spending, not only because the economy can steam along by itself, but because extra government spending could be inflationary—too much demand chasing too few goods will send prices up. That's the notion. The fact is that even though the economy has gone through many up-and-down cycles since 1929, the government has run a budget surplus in only eight of those years, most recently in 1969.

This seesaw between budget deficits and surpluses brings us to the revival of an idea that has been out of fashion for several decades, the *balanced budget*. The notion is simplicity itself: The government should not spend more or less than it receives, but exactly what it receives. Supporters of a bal-

anced budget argue that it's the only responsible way for the government to act. Critics reply that if the federal government committed itself to running a balanced budget, then it wouldn't be able to soften the impact of economic slumps. See DEFICIT SPENDING OR FINANCING.

Business Cycle

The technical term for the ups and downs in the economic growth rate. Although the word "cycle" implies some regular, rhythmic movement, business cycles are erratic. They encompass varying periods of time—the ups have ranged anywhere from nine months to eight years, the downs from half a year to six years—and they vary in severity and plenitude. Their only true cyclical aspect is that ups are followed by downs which are, in turn, followed by ups.

Business cycle chat employs some specific terms: The lowest point, which occurs during a recession (or depression), is called a trough; the highest point, which occurs during a boom, is called a peak. The in-between periods are called expansions or contractions.

The economic statistic used to track business cycles is the trend (or historical) growth rate of gross national product, minus inflation. When GNP falls below its trend, economists say that the economy is contracting; when it grows above its trend, the economy is expanding. The long-term trend of GNP, by the way, goes gently upward—a reflection of the fact that the periods of expansion have been longer and stronger than periods of contraction.

Ideally, of course, there would be no downs. Ideally, economists and politicians would be able to manipulate the economy to ensure steady growth without reversals. Realistically, however, the economy is not all that manageable.

Most obviously, it is prey to events beyond the control of economists. These types of events are called, appropriately enough, shocks. Inflationary shocks, for example, can be particularly menacing to economists' dreams of stable prices.

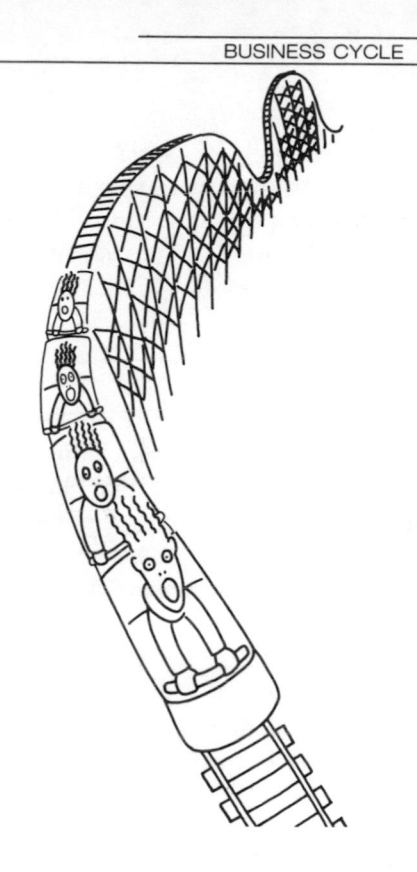

BUSINESS CYCLE

Consider the weather, a traditional wild card: Floods or droughts can translate into a bad harvest which, in turn, means suddenly high food prices. Or consider the severe inflationary shocks in the 1970s when OPEC decided to raise oil prices dramatically.

Almost as obviously, the economy is also prey to events orchestrated by its players. A giant tax increase, for example, might curtail spending severely, causing economic activity to sicken. Or, optimistic firms might produce too much, causing an "inventory slowdown."

The business cycle, then, is likely to remain one of those exasperating facts of economic life, and good times will follow bad, bad times will follow good. . . . See INVENTORY.

Cc

Capital

Any physical item used to produce things. Capital covers a broad range: from humble implements like hoes and hammers to huge factories, complicated machinery, and sophisticated equipment.

As most introductory textbooks will tell you, capital is one of the three factors of production; the other two are land and labor. A classic example of this trio describes how a garden plot (land), a farmer (labor), and a hoe (capital) are combined to produce tomatoes. The hoe is important in the production process because it enhances the productivity of land and labor: Sure, the farmer could scratch out a tomato garden without a hoe, but he can produce more tomatoes, in the same amount of time, using a hoe.

The act of producing capital goods is itself a form of investment because it uses resources to make goods that are not consumable, but are used to make other goods (machine tools, for example). Hence the origin of the phrase "seed capital" to describe an economy's investment in future productivity. If an economy consumes its seed capital, it diminishes its chance to grow in the future. Thus, the amount of capital goods—or, the size of the capital stock—in an economy is an important measure of viability.

Like any other investment, the value of capital goods depends on the income they generate in the future. The value of the farmer's hoe, for example, is not its purchase price, but how much income the hoe will earn during its useful life, discounted by the interest rate or the carrying cost of the hoe.

Capital Formation

The net addition to the capital stock in a given period. Capital formation is positive when expenditures on new goods exceed the consumption of old ones.

Both capital expenditure and consumption go on all the time. Businesses, especially growing ones, add new capital

when they expand production facilities, update equipment or replace worn-out machinery. Businesses also run down existing capital when they use it for production (depreciation) or junk it entirely. What economists call a positive rate of capital formation—a good sign for the economy—means that capital stock is being added faster than it is depreciating.

Capital Market

The financial market for long-term investment and savings. Just like other markets, the capital market serves to link up buyers and sellers. In this case, sellers are individuals, businesses, and governments wishing to raise funds (capital) and buyers are individuals, businesses, and governments wishing to save money.

The instruments through which this linkage takes place are varied. They include corporate stocks and bonds; residential, commercial, and farm mortgages; federal, state, and local government bonds; and even consumer and business loans. These instruments also vary in the length of time for which the loan is made—from a year to several decades—making the capital market distinct from the money market where instruments are very short-term, sometimes overnight.

A healthy capital market—one with lots of buying and selling of many different instruments—is essential to a healthy economy.

Take, for example, a corporation wanting to build a new factory. First, it would be mighty unusual to have the necessary cash on hand, so the money must be borrowed. Second, the company wants to make the best use of its resources, so it needs flexibility in the way it borrows. Third, such a construction project takes a long time, so the company needs to be sure that the money won't run out before the project is completed. And fourth, lenders want to be assured that their investment is liquid (somebody will buy the company's loan instrument if need be), so they want a market with lots of

buyers. In other words, the corporation needs a market in which a large chunk of money can be raised, in a number of ways, for a long-term investment.

Just as corporations need an efficient capital market, so does the economy. Absent a capital market, corporations might not expand production, thus they would not create new jobs and so on and so forth. See CAPITAL.

Cartel

A group of otherwise independent firms that band together in order to control their industry, usually by cutting back production to raise prices.

Their motivation is straightforward: to transform a competitive situation (many producers making decisions individually) into a monopoly (many producers making decisions as if they were one entity). Cartels can operate in several ways, but for a general idea of how they work, consider the (in)famous Organization of Petroleum Exporting Countries (OPEC). The oil industry is relatively easy to organize—or, to use the buzzword, cartelize. While only a handful of countries in the world are endowed with the natural resources to produce oil, almost every country in the world needs oil. In other words, there are few sellers and lots of needy buyers.

Before OPEC, each oil-producing country churned up as much as it could and competed with one another on prices. After OPEC was formed, however, most oil-producing countries agreed to reduce their individual output in order to achieve new, higher prices. When OPEC got going in 1973, the price of oil was about $3 a barrel. But with OPEC setting production quotas for each of its members, the price rose to over $30 a barrel within a decade. Not bad.

But OPEC has been subject to one of the classic problems encountered by cartels—being killed by its own success. High prices create a strong incentive for members to cheat on the cartel by producing more than the agreed amount. The

members figure they can increase their own revenues by selling more output at the higher prices. As these sneaky members increase production, however, they bring more output onto the market which, of course, pushes prices back down. And so, when some OPEC members gave in to the temptation to cheat on the cartel, they helped cause steep price declines. See MONOPOLY.

Ceteris Paribus

A handy Latin phrase meaning "all other things being equal." Economists find ceteris paribus indispensable for pronouncing certain laws of economic behavior. (To sit through a semester of introductory economics is to hear that little phrase repeated endlessly.)

Take the law of the downward-sloping curve, for example. If the price of a good falls, then more of the good will be bought—ceteris paribus. That is, if tastes, income, and prices of other goods are held constant. However, if there is no ceteris paribus, and taste, income, or prices of other goods are free to change, then the tidy law of downward sloping demand becomes very messy to explain. Say the price of quiche goes down but, at the same time, quiche ceases to be a fashionable dish. Then less quiche, not more, will be bought. (You can see why the utterance of "ceteris paribus" is so handy and so frequent.)

Chicago School

Informal designation for economists who are either thoroughgoing believers in the virtues of the free market and/or monetarism. Membership is not confined to an actual presence at the University of Chicago (Milton Friedman and George Stigler do teach at Chicago while Karl Brunner, Allan Meltzer, and Beryl Sprinkel do not.) The derivative term "Chicago Boys" is usually employed by nonbelievers. See MONETARISM.

Classical Economics

A school of thought that emphasizes free markets, competition, private initiative, and small government. The roll of classical economists includes some very big names, beginning with Adam Smith and followed by Jeremy Bentham, David Ricardo, Thomas Malthus, John Stuart Mill, Jean Baptiste Say, and Alfred Marshall. Classical economics dominated thinking for almost two centuries until it was eclipsed by John Maynard Keynes in the 1930s. See SMITH, ADAM.

Comparative Advantage

A standard notion in international trade talk meaning that every country, no matter how lowly, can produce something for less *relative* cost than any other country. It explains why a superior country—say, one that can produce almost everything—specializes in exporting only a few things.

The principle of comparative advantage was first enunciated by the British economist David Ricardo (1772–1823). Ricardo's example, involving Portugal and England, wine and cloth, still lives as the standard textbook explanation. Briefly, the Ricardo illustration assumes that both countries make both products. Portugal, however, is able to produce wine and textiles cheaper than England. It takes Portugal 80 labor hours to make one unit of wine, England 120 hours; and it takes Portugal 90 labor hours to make one unit of cloth, England 100 hours. Using trade lingo, Portugal is said to have an absolute advantage in making wine and cloth.

At first blush, then, it looks as though Portugal is better off making its own wine and cloth. However, now consider that one unit of Portuguese wine can only "buy" eight-ninths of a unit of Portuguese cloth (80/90 hours) while one unit of Portuguese wine can "buy" one and one-fifth units of English textiles (100/80). Similarly, one unit of Portuguese cloth can buy one and one-eighth units of Portuguese wine, but only one and one-ninth units of English cloth. Obviously, the best deal for Portugal is to trade its wine for English cloth—that's

the arrangement that nets Portugal the most units. Again, in trade lingo, Portugal is said to have a comparative advantage in wine.

But the example doesn't end quite yet. Here's the point: Since Portuguese wine buys relatively more English textiles than anything else, it makes sense for Portugal to specialize in making wine and exporting it to England. Thus—if your eyes are still tracking—comparative advantage leads countries to specialize in their lowest relative cost products and trade among each other.

Competition
A favorite word for economists, and one which has a very specific definition. When used by itself, competition refers to a pure situation—more formally called "perfect competition." Here, there are four conditions necessary to produce a perfectly competitive market structure:

1. A marketplace made up of many buyers and sellers, each buying or selling just a tiny fraction of the particular product being bought and sold.

2. A product that is homogeneous: that is, the product sold by any one firm is indistinguishable from the products sold by other firms.

3. An economic situation that permits would-be sellers to set up shop, and current sellers to close up shop, with relative ease.

4. Good information and communication so that all buyers and sellers are informed about prices and sources of supply.

Consider, for example, the soybean industry, which is very close to being perfectly competitive. Millions of farmers produce and sell soy to millions of buyers, the soy is pretty much the same no matter who grows it, farmers can grow soy one year and wheat the next, and information on prices is easily attainable.

Taken together, all these conditions create a situation in

which nobody has control over prices. Both sellers and buyers must do business at whatever price is determined by the market, the price that results when all the farmers and all the consumers get together. A farmer cannot charge more for her soybeans than her neighbor does and still find willing buyers, and a buyer cannot offer less than other buyers do and still find willing sellers.

Competition, as outlined above, is hardly a situation jam-packed with excitement and intrigue. No clever advertising campaigns for Brand X soy, no corporate battles between Farmers A and B. Indeed, one might wonder how the notion of competition came to be so firmly associated with capitalism, with firms slugging it out and rewards going to the toughest or brightest.

Well, the answer is that perfect competition is hardly the way of the real world. Capitalism is more correctly associated with the terms "imperfect competition" or "monopolistic competition," terms that describe market structures that do not meet those four, somewhat ideal, conditions.

Typically, for example, sellers can distinguish their products by using packaging and advertising to convince buyers that Brand X beer is not only different from Brand Y, but better—higher quality, snappier taste, or preferred by glamorous people. Sellers who are able to differentiate their products can, in turn, exercise some control over prices. Beer drinkers might be willing to pay more for a brew consumed by hunky he-men. See MONOPOLY.

Conglomerate

A huge corporation consisting of many different types of businesses. It's not unusual, for instance, to have a conglomerate that bakes bread, runs hotels, flies airplanes, oversees a telecommunications business, and sells insurance. Conglomerates don't spring full-blown into the pages of the *Forbes* 500s, however. They are the result of a series of mergers or

takeovers, usually undertaken by firms wishing to diversify their risk: If profits slip in one market, the conglomerate can still count on profits from businesses operating in other markets.

Conspicuous Consumption
The idea that people purchase things primarily to impress their neighbors.

Conspicuous consumption was first enunciated by Thorstein Veblen in *The Theory of the Leisure Class* (1899). Veblen saw evidence of conspicuous consumption everywhere, by rich and poor alike. Indeed, he even classified women, who could be decorated with jewels and furs to demonstrate a man's success, as conspicuous consumption. Conspicuous consumption has become an important assumption in assessments of consumer psychology: Advertisers use it shamelessly and successfully to promote their products.

Consumer Goods
Any manufactured item used by individuals as opposed to businesses.

Consumer goods come in two categories. Nondurable goods means stuff that doesn't last over three years, like shoes, cheeseburgers, and cigars. Durable goods means stuff that should last longer, like cars and toaster ovens. People usually spend three to four times more money on nondurables than durables.

Consumer Price Index
Monthly measure that tracks changes in the price level, also known as the CPI. (This is usually what people refer to when they talk about the cost of living going up—or, rarely, down.) The CPI was hardly a big deal when it began in 1917. Very

little depended on it, so very few people paid any attention to it. Now, however, the CPI is one of the most intensely watched government measures—especially during periods of inflation. And with good reason. A lot depends on its every wiggle: it determines the size of payments under all sorts of contracts, from labor and royalty agreements to Social Security checks. A small bounce can trigger the payout of billions of extra bucks.

Just as its importance has brought notoriety, it has also brought complaints about its usefulness. Green-eyeshade critics, for example, complain that the CPI does not accurately measure the real world. Their criticism is aimed at the way the CPI is constructed:

The CPI records changes in the cost of living by measuring the prices of typical items bought by typical families, colorfully called the typical "market basket." The market basket is divided into categories like food and drink, housing, clothing, transportation, medical care, entertainment, gasoline, and so forth. These items are also supposed to appear in the market basket in the same proportion that they appear in the family budget: say, 40 percent for housing, 5 percent for medical care, and .001 percent for peanut butter. Thus, even if the price of peanut butter rockets up, if the price of housing droops a bit, the CPI will record only a slight rise in the cost of living.

All goes well if people's market baskets stay the same, year after year. But here's the problem: People are forever altering the contents of their baskets.

Consider what happens when prices change. If peanut butter becomes wildly expensive, consumers will stop buying it and purchase, instead, apple butter. But if peanut butter is in the market basket and apple butter is excluded, peanut butter prices will move the CPI upward even though people's actual cost of living remains the same. So, too, tastes change for reasons other than prices. If consumers become terrifically health-conscious, the percentage of family budgets going for medical care will increase. But, again, the CPI will

understate changes in the cost of living because it continues to weight medical care at a fixed 5 percent. Thus, critics say, the CPI, with its unchanging market basket, misrepresents actual changes in the cost of living.

Consumer Sovereignty

The power of consumers in a free market system to determine what gets produced.

This notion, like most theoretical constructions, sounds irresistible. Consumers use their dollars to vote on goods and services, thus dictating their demands to producers. If consumers like something, for example, they will vote lots of dollars for it. The big-dollar vote gives the makers of that product lots of dollar profits. The big-dollar profits, in turn, lure other producers into the business. There are, of course, some nontheoretical objections to consumer sovereignty: The most damning is that producers can organize themselves into monopolies and dictate to consumers.

CONSUMER SOVEREIGNTY

Consumption or Consumption Spending
The total money spent on goods and services.

People have two choices when it comes to deciding what to do with their money: They can spend it or save it. While decisions on what they spend it on depend on individual tastes—one person's diamond tiara is another person's vacation in France—decisions on the amount spent (and, thus, on the amount saved) depend on two identifiable and generalizable variables.

The more important of these variables is income. John Maynard Keynes first formalized the relationship between consumption and income into something called the con-

CONSUMPTION SPENDING

sumption function. This relationship simply says that as disposable income rises, expenditures will, too, though by less than the increase in income because some money will be saved. Consumption is also a function of wealth—the richer people are, the more they will spend.

Beyond these two relationships, there are several theories about what determines spending over the long term. The life-cycle hypothesis says that spending is uneven. At the beginning and end of the life cycle, when people are starting out or retired, income is low; thus people spend more than they save. In the middle of the life cycle, however, when income is higher, people save more than they spend.

A second theory, called the permanent income hypothesis, says that people save and spend a constant percentage of income, no matter where they are in the life cycle. In other words, a bout of unemployment will not cut spending dramatically.

And a third theory maintains that consumption spending is based on perceptions of relative income. That is, those with incomes below the national average spend more, proportionately, than those with incomes above the average.

Whatever determines individual trade-offs between consumption and savings, consumption spending in aggregate is an important determinate of economic growth. Indeed, it constitutes almost two-thirds of GNP. Thus, as a general rule, when consumption spending is strong, so is the economy.

But general rules cry out to be broken, and this one is no exception. If the flip side of strong spending means weak saving—people are spending at the expense of saving—then the economy will eventually falter. It's impossible to maintain growth if the economy isn't making the necessary investment in the capital stock that underwrites growth. Also, too much spending can be inflationary if the desire to spend bumps up against the limits of the economy to produce.

Consumption spending, then, is like eating. It is strictly required—and, goodness knows, it's fun—but too much of it will lead to unhappy results.

Cost-Benefit Analysis

A technique used to determine whether the benefits to society of undertaking a public project outweigh the costs.

Say the government is considering the construction of a new airport. Cost-benefit analysis totes up the costs of the land, the necessary construction, the noise pollution from air traffic, even the loss of the scenic beauty, and so forth. Those costs are then compared to the benefits from things like decreased traveling time and less congestion at existing airports.

Critics of cost-benefit analysis complain that some of the categories are too nebulous—for example, it's hard to attach a dollar value to noise pollution. Too, they argue that cost-benefit calculations are a waste of time because government decisions are political, not strictly economic. Supporters reply that cost-benefit analysis is an important way for decision makers to get an idea of a project's worth, and that some information is better than none.

Cost-Push Inflation

A general rise in the price level coming from the cost, or supply side. See INFLATION.

Council of Economic Advisers

Established in 1946 to advise the president on economic matters and inform Congress on the health of the nation. Its chief public function is to make forecasts of the country's economic growth—forecasts that usually undergo major revisions by their midyear "updates."

The council is usually made up of three economists, one of whom is the chief, and a staff of other economists. It reached its peak of prestige and power during the sixties when its advice on managing the economy was not only listened to, but occasionally followed. Back then, the ambition of academic economists who wished to be movers and shakers of public policy was to be "on the council." The failures of eco-

nomic policy in the seventies, however, seem to have soured the whole enterprise, and the brightest economists now prefer to stay at home.

Credit
The mechanism that allows goods or services to be sold on the promise of future payment.

It's hardly hyperbole to say that credit lets the world go round. Business uses credit to finance everyday matters (like inventories) and occasional undertakings (like plant expansions). And consumers use credit to finance spur-of-the-moment purchases (baubles), and well-thought-out ones (cars and houses).

"Buying on credit" can have a sort of immoral ring—is it right for people to obtain goods or services without paying for them? Well, credit is really a kind of substitute money. So, the quick response to any moral objection is that buying on credit is essentially the same as buying for cash, especially since real coin follows the credit transaction. (In fact, in the vast majority of credit transactions, the piper is eventually paid.) Beyond moral considerations, the fact is that a credit economy functions with greater smoothness and at a higher level than a strict cash economy.

First, credit is more convenient than money for big purchases. Without credit, businesses would have to employ fleets of armored cars to ferry their cash around, to and from suppliers and to and from bankers. Ditto for consumers: Without credit, shopping sprees would be incredibly burdensome to mount. To put it in economists' lingo, credit allows people to economize on their use of money.

Second, credit facilitates a multitude of activities by allowing people to have the use of things before they actually own them. Installment credit, for example, lets a homemaker pay for the service provided by a washer-dryer—and process tons of laundry—before he owns the appliance outright. It's just like making mortgage payments on a house: The mortgagee

can live in the house while paying for it and, after enough payments have been made, she can then take possession of it. In other words, people don't have to postpone the benefits from a pricey item just because they can't pay the entire cost up front.

Credit Crunch
A situation in which people want more credit than the economy can easily make available.

Credit crunches usually happen after the economy has been growing at a rapid clip—that is, after a period during

CREDIT CRUNCH

which business has borrowed a lot of money to expand operations and consumers have taken on a lot of debt to finance purchases. This strong demand for credit draws down the pool of credit (the savings made available to spenders), leaving would-be spenders to battle over the remaining funds. The crunch itself comes when the battle drives up interest rates (the price of credit). This type of credit crunch is considered a normal stage in the business cycle.

Credit crunches can also be manufactured by the Federal Reserve Bank. For example, the Fed can suddenly and severely restrict the supply of money in the economy, which then drives up interest rates. Typically, the Fed engineers a credit crunch to cool off an overheated economy—a situation that it fears will either lead to, or has already caused, high inflation.

In either case, when the credit crunch drives interest rates high enough, businesses and consumers cut back on their spending, thus sending the economy into the contractionary stage of the business cycle. See BUSINESS CYCLE; FEDERAL RESERVE SYSTEM.

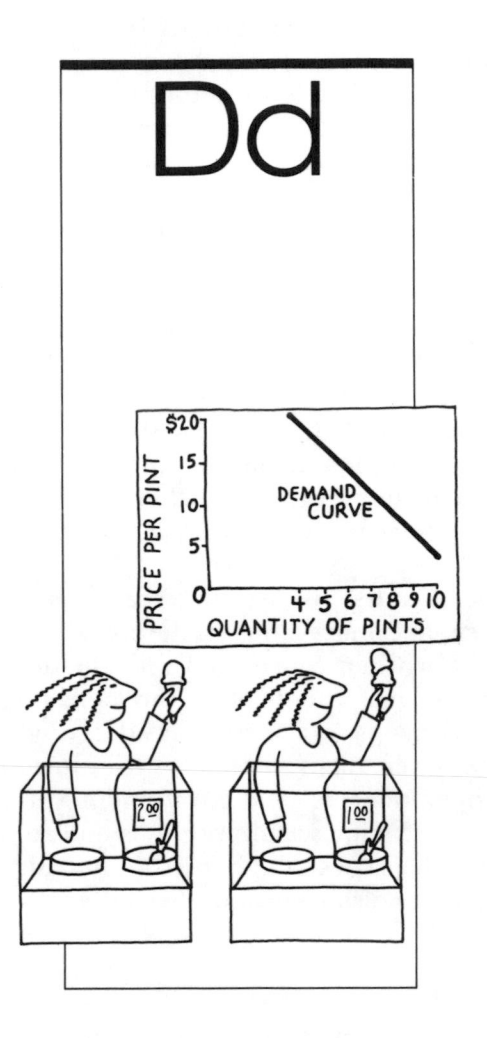

Debt Limit

A legislated ceiling on the amount of outstanding debt that a government can incur.

The purpose of debt limitations is to hold government spending within reasonable bounds of its current revenues. The stricture is based both on historical experience and common sense: government incentives are all on the spending side and not on the taxing side; if left unchecked, government debt would merrily mount up.

Debt limitations have real bite at the state and local levels. Most state governments operate under debt limits mandated by their constitutions and most municipalities have debt limits set by state governments. The federal government also has a debt ceiling, but in this case, the limit is, well, flexible in the extreme.

Congress sets the debt ceiling for the federal government. So when the government bumps up against that ceiling—which it does in almost every year there's a budget deficit—it must ask Congress to raise the limit. What follows is pure circus.

Congress, of course, is directly responsible for the budget deficit. (After all, Congress directs the spending and legislates the taxing.) But when confronted with the debts it has created, Congress turns coy. It often threatens not to raise the limit. Congresspeople make speeches decrying irresponsible deficits. The government bureaucracy, in turn, threatens to shut down operations, to send government workers home, to stop issuing checks for social programs and so forth. In the end, Congress legislates a higher debt limit and the government continues to disburse funds. See BUDGET.

Deficit Spending or Financing

When the government spends more money than it takes in, the difference is covered—or financed—by government borrowing.

Deficit spending, which is a deliberate policy, has three excuses.

One, deficit spending is undertaken when the government desires to stimulate the economy. Here, the government acts when the business cycle hits its drag phase: Deficit spending provides oomph to the economy by injecting purchasing power that is not forthcoming from the private sector. Government spending is supposed to rev up the economy, pushing it into the expansion phase of the cycle. (In this sense, deficit financing should only occur during the down part of the business cycle: During the expansion phase, the government should run a budget surplus to avoid overheating the economy.)

Two, deficit spending is necessary during war. Here the government spends whatever it thinks it must in order to buy military victory.

Three, deficit spending is a way of life. Here, government spending has proved seductive for Congresspeople who spend whatever they think it'll take to buy political victory. See BUDGET; BUSINESS CYCLE.

Deflation
A situation of generally falling prices.

Deflation, like all big economic events, can be caused by many things. Usually, however, it's the result of high interest rates that have forced business and consumers to cut back on their spending.

As an idea, deflation doesn't sound like a bad thing, especially if it follows a period of inflation. Indeed, the very event of falling prices should call forth its own solution: Falling— or lower—prices will encourage people to start spending again. The problem is that this only holds true in theory. In reality, deflation can be quite a bad thing because, perversely, it discourages spending.

First, all prices do not fall during a period of deflation:

Some prices are maintained by the government (farm prices), and some are set by long-term agreements (labor contracts). When high interest rates cause a drop in spending, some businesses will be left with too much output, or more stuff than the market will buy. Now, if firms simply lowered their prices, people would start buying again and the process would be self-correcting. But many firms cannot lower prices; if, for example, they are constrained by labor contracts, then a price cut means they will not be able to cover their fixed costs.

This situation, known as sticky wages, causes firms to cut back output rather than reduce prices. Of course, if they produce less, they will need fewer workers. So, one result of deflation can be unemployment. And unemployment means less spending as people are laid off.

Second, consider what happens in the sectors of the economy where prices do fall. Owners of these businesses will earn less and, quite probably, spend less. And so, a second result of deflation is even less spending.

Thus, deflation can have a kind of ripple effect: If prices fall, people earn less; if prices can't fall, people will be unemployed. Both will cause spending to drop further and the economy to languish. In the real world, then, deflation can be a very bad thing. That's why government undertakes various fiscal and monetary policies, like deficit spending or increasing the money supply, to stabilize prices.

There is one possible bright note for an economy undergoing a deflationary period. When prices decrease at home, but spending remains strong abroad, foreign demand for domestic output will increase. If foreign spending is strong enough, the economy can get itself going again without government intervention.

(Deflation should be distinguished from something called disinflation. Unlike de-flation, dis-inflation is a good thing; it means that prices are not increasing quite as fast as they were, or that inflation is slowing down.) See DISINFLATION; INFLATION.

Demand

Shorthand expression for the relationship between the desire to buy something and its price. It's hard to underrate the idea of demand in economic analysis: Demand—and its twin, supply—are two of the foundations on which modern economics is built. (Giving rise to the joke that if you teach a parrot to say "supply and demand," you will have created an economist.)

Demand has a very exact, mathematical meaning for economists. Learning that precise meaning involves plowing through several chapters in an introductory textbook, absorbing ideas like indifference curves, marginal utility, and budget constraints, and—at a more sophisticated level—understanding the rules of differential calculus. In the end, though, both the tutored and the untutored will come out the same, commonsense place: People will almost always demand more of a good when its price falls, and they will demand less of it when its price rises.

The relationship between the quantity demanded and its price is best expressed graphically by something called a demand curve. Consider a demand curve showing how many pints of cherry vanilla ice cream an ice-cream maven will demand over a month's time. The possible prices per pint can be aligned along the vertical axis (from $4 to $10). The number of pints along the horizontal axis (from zero to 20). The demand curve consists of plotting how many pints the maven will buy at each price: Say, for example, at $10 a pint, the maven will demand eight pints, at $9 a pint, she will demand nine. When the points showing how many pints will be demanded at each price are connected by a line, that line will slope downward from left to right. That is, there is an inverse relationship between the quantity demanded and the price: As prices fall, quantity increases. The line's downward direction, what economists call its negative slope, is due to the so-called law of demand. (When properly drawn, the line is actually a curve because of the elasticity of demand.)

What's behind the law of demand? Beyond common sense

—that the cheaper something is, the more of it people will buy—economists have identified two (fancier) reasons to explain why the demand curve goes down: the income effect and the substitution effect.

The income effect reflects the fact that as the price of cherry vanilla falls, consumers will gain more purchasing power—they can buy more ice cream per dollar than before. And the effect on their behavior is the same as if their real income had increased: Consumers are richer, so they can afford to—and will—buy more ice cream.

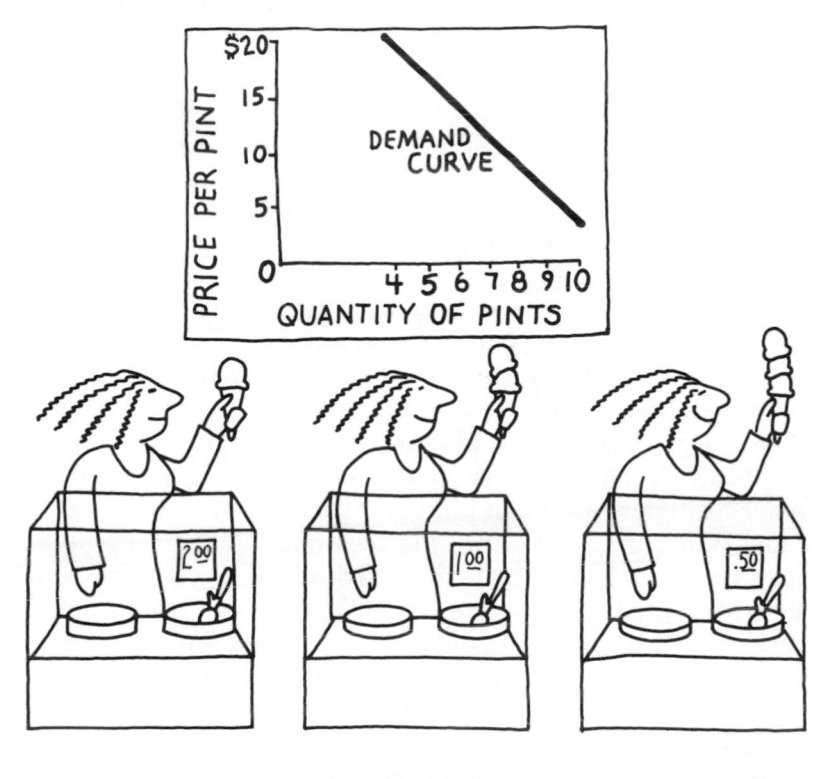

DEMAND

The substitution effect reflects the fact that as the price of ice cream falls, other sweets will become more expensive relative to ice cream. The new cheapness of ice cream gives consumers a strong incentive to buy more ice cream at the expense of fudge cookies. They substitute purchases of the cheaper goods for purchases of the pricier goods.

There is a third reason that demand curves slope downward. As prices fall, people who weren't previously consumers—who were priced out of the market—can suddenly afford to buy. As these new consumers come into the market, the quantities purchased will increase accordingly.

It's possible to take consumer behavior all together (as in reason three) into one giant demand curve. Possible but a little tricky. Obviously, all consumers are not alike: At $10 a pint, some will demand four pints of ice cream, others won't want any; at $15 a pint, some consumers will demand eight, others will be content with two. Nonetheless, it's possible to add together all the individual demand curves for cherry vanilla and get something called the market demand curve.

Although ice-cream makers routinely estimate the market demand for their product at various prices—and set their prices and production accordingly—market demand can be slippery. Several events can upset the relationship between price and quantity demanded.

First, consumers' tastes or preferences could change: In the case of ice cream, people might decide that fudge cookies are more tempting. Second, consumers' income could change: A drop in income might make people feel that ice cream, because it's a luxury good, is dispensable. Third, the population could change: The number of possible consumers at prime ice-cream-eating age might decrease. And fourth, prices for close substitutes could change: A decrease in the price of fudge cookies might cause consumers to buy more fudge cookies.

Any of these changes will cause the quantity of ice cream demanded to fall across the board. That is, at each possible price ($10, $9, $8, $7, and so forth), the market will demand

500,000 fewer pints than previously. And ice-cream makers will suddenly discover that they sell fewer pints at any given price. When the entire relationship between price and quantity changes like this, economists say the demand curve has shifted down.

So far, the analysis of demand had been satisfyingly tidy. There are, however, two loose ends. Demand for some goods or services can increase when prices go up, while demand for other goods or services can decrease when prices fall.

The former category is likely to include already pricey, frill-type items, like Italian perfume, designer fur coats, and racy autos. Here, higher prices call forth more buyers either because high price itself confers "status" on the buyer, or because high price is taken as a sign of higher quality. The latter category is made up of cheaper, nonfrivolous goods, like potatoes. Here, lower prices free consumers' income up to buy other goods, so they buy less of the original good. These exceptions to the law of demand are rare, and economists don't like to admit they exist at all. See ELASTICITY; SUPPLY.

Demand-Pull Inflation
General increase in the price level when demand outruns available supply. See INFLATION.

Depreciation
A fall in the price of a currency relative to other currencies. Under a system of floating exchange rates, the market sets the value of currencies. A nation with a "weak" currency— one that other nations don't want to own—will see its currency decline in value relative to other currencies. The nation with the depreciating currency usually finds itself with a trade surplus because its exports become cheaper and its imports become more expensive. See BALANCE OF PAYMENTS; FOREIGN EXCHANGE.

Depression

When economic growth turns steeply negative for a pro-longed period. During the Great Depression of the 1930s, for example, employment, production, prices, wages, interest rates, and profits zoomed down and stayed low.

In business cycle jargon, depressions are just like recessions, only more so. (Recessions are defined as any period when gross national product falls for at least two quarters.) However, the party in power prefers to call any downturn in the business cycle a recession; so depression has become a political term used by the party not in power. See BUSINESS CYCLE; RECESSION.

Deregulation

Umbrella term to describe the dismantling of various laws that regulate economy activity.

Obviously a whole lot of regulating had to go on to reach the point where:

1. Deregulating became the necessary next step.
2. The dismantling process itself was important enough to deserve a special term.

Well, beginning in the 1930s, a whole lot of regulating did go on.

Granted, governments find meddling with the economy irresistible. But the laws and regulations that got their start under the New Deal were pretty ferocious: The Roosevelt administration carved out giant chunks of the economy and brought them under the purview of government agencies. Once the effort began, it—like all bureaucratic endeavors—multiplied. Today, tens of thousands of bureaucrats spend billions of dollars in an effort to direct economic activity. The cost in tax dollars alone is sufficient to draw attention.

But the dissatisfaction with regulation goes deeper—and is more informed—than just tax money. All this activity has not been successful in promoting a healthy economy. In fact, there are tons of evidence that the costs of regulation outweigh the benefits.

Essentially, the laws and regulations that have been un-done restricted competition. Airlines, railroads, truckers, tele-phone companies, banks, and brokerage houses had all been granted little monopolies over their businesses. Thus, these firms had few incentives to perform efficiently and keep their costs down—costs could always be passed on to consumers in the form of higher prices.

This lack of competition created megaproblems. Along with forcing consumers to pay higher prices and limiting their choices in products and services, economic growth was slower and inflation higher because inefficient firms were sheltered from the rigors of the free market. And this isn't just purple prose: Regulation's deadening impact can be gauged by the revitalization that deregulation has brought.

In three of the industries that have been undergoing deregulation—financial services, telecommunications, and transportation—consumers have been treated to falling prices and an expanding array of products and services. Too, the industries themselves are exhibiting signs of vitality: New firms are starting up, of course, some old firms are closing. Although the direct impact on the economy is harder to call, deregulation has already brought diminished inflation and greater productivity.

One last note, the deregulation that has been going on is economic as opposed to social. That is, the government con-tinues to make laws and regulations bearing on safety, health, and the environment. See ANTITRUST; MONOPOLY; REGULA-TION.

Devaluation
Lowering the value of a nation's currency under fixed ex-change rates.

A fixed exchange rate is what happens when governments declare that their currencies will have a fixed value in relation to other currencies or to gold. Devaluation, then, can only

come about as a deliberate action on the part of government.

Say, for example, $35 in U.S. currency buys one ounce of gold. Now, if the United States wants to devalue, it will raise the price of gold by declaring that henceforth it will take more dollars to buy one ounce of gold. If, for example, the United States raises the price of gold to $38, then it has lowered the value of the dollar by 7.9 percent. That is, it now takes 7.9 percent more dollars to buy one ounce of gold. If other nations keep the price of gold in their own currencies constant, then the value of the dollar will have fallen relative to them. Simply put, it will take more dollars to buy a French franc or a German deutsche mark.

Devaluation is usually undertaken when a nation is running a troublesome trade deficit with other nations. By decreasing the value of its currency—and thus increasing the value of other currencies—the devaluing nation makes it cheaper for other nations to buy its goods and more expensive for its own citizens to buy foreign goods. If all goes well, the value of exports from the devaluing nation will go up, imports will go down, and the trade deficit will disappear. The United States went off a fixed exchange rate in 1971, allowing the dollar to float up or down as the foreign exchange market dictates. See BALANCE OF PAYMENTS; FOREIGN EXCHANGE; GOLD STANDARD.

Diminishing Returns

A heavy-duty principle of economics. It states that the continued addition of one factor of production—land, labor, or capital—when the other two are held constant will eventually lead to less and less additional production.

This principle is best explained through an example, a simple example. Say you (labor) own a potato patch (land) in which you have planted potato seeds (capital). And let's say that labor is the easiest factor of production to vary.

If you don't spend any time farming your potato garden,

you won't harvest any potatoes. (Zero labor input yields zero output.) If you spend one day in your potato garden, you can expect output to be five bushels. If you ask a friend to help, output climbs to nine. The marginal, or extra, increase in output from the second unit of labor is thus four bushels. A third friend might yield fifteen bushels, for a marginal increase of six bushels. But let's say that when the fourth friend joins in, total output increases to twenty, yet—as you can see —marginal output falls back to four.

In other words, total output increases as you add labor, but eventually the marginal increases in output start to fall. Why? Well, since the other two factors don't change—your garden plot and the seed you have planted stay the same size—you and your friends will start tripping over one another. Everybody becomes less productive.

The law of diminishing marginal returns is very important for firms that have to hire labor or buy land or capital. It helps them figure out how much of the three factors of production to use and how much output will put them at their profit-maximizing point.

Disequilibrium
A market state directly opposite of equilibrium.

The market is said to be in disequilibrium when either of two things happen: Prices are so low that there are too many buyers and not enough sellers, causing shortages; or prices are so high that there are too many sellers and not enough buyers, causing surpluses.

Usually the market will straighten itself out and achieve equilibrium. In the case of shortages, the abundance of buyers will cue suppliers to raise prices; in the case of surpluses, the scarcity of buyers will cue suppliers to lower prices. In other words, disequilibrium is an unstable condition. Market forces will push prices toward stable levels. See EQUILIBRIUM.

Disinflation

A slowing in the rate of inflation. Granted that sounds disingenuous, but it's not. Say prices rise 7 percent a year for several years, putting inflation at 7 percent. But if the price rise suddenly drops to 4 percent a year, then—even though prices are still going in an inflationary direction—the decline in the rate is disinflationary. Don't confuse disinflation with deflation, where prices actually go down. See DEFLATION.

Disposable Income

Take-home pay, or what's left after you pay income and Social Security taxes.

After you take it home, disposable income can be saved (in which case it's called savings) or spent (in which case it's called spending). Obviously, the size of disposable income is important to the receiver. But it's also crucial for the economy—and for the economists whose business it is to forecast the course of the economy.

People spend most of their disposable income, historically about 90 percent. And this spending constitutes a mighty engine for economic growth. In fact, the biggest component of gross national product is "personal consumption expenditures," which account for almost 65 percent of GNP. Thus, when disposable income is growing, it's a sure bet that GNP will grow. Conversely, when disposable income is flat or declining, the economy will suffer. Not only will spending languish, but so will saving—and less savings and investment today means weaker growth down the road.

Dissaving

Loosely, the opposite of saving; more rigorously, when spending exceeds income over a certain period—with the difference financed either by borrowing or by drawing on past savings.

DISSAVING

There are many reasons for dissaving. It can be temporary, due to the purchase of a big-ticket item, a heavy unexpected expense, or a spell of unemployment. Or, it can be rather permanent, due to normal retirement or abnormal profligacy. Whatever the cause, the result is the same: Dissavers run down their assets.

Dumping

A rather inelegant name for the practice of selling products abroad at cheaper prices than they are sold at home.

Dumping can be prompted purely by internal business considerations. Firms might want to get rid of stuff that cannot or will not be absorbed at home—either because the products don't have a ready market or the sale is illegal for health or safety reasons.

Dumping can also be part of government policy. Firms

might be encouraged to dump their goods when their country is running a trade deficit. Here, the government subsidizes companies—pays them to produce and export—in order to restore the balance of trade.

Dumping is outlawed under the General Agreement on Tariffs and Trade, which most major nations have signed. GATT gives offended governments the power to set tariffs against offending dumpers. (Tariffs, which are taxes on imports, thus raise the price of the dumped goods.) This is usually a major event, however, because dumping is difficult to prove.

Despite its name, dumping is not a totally undesirable act. The dumper gets to off-load merchandise, and the dumpee gets to buy the stuff for cheap. The competing firms in the dumpee economy, however, don't quite see things this way. When dumped goods are sold below domestic prices, these firms will lose sales. And, if the dumping continues for any length of time, firms will have to start laying off workers. That's usually when the government starts complaining and threatening to slap a tariff on the imported goods.

Durable Good
Just as it sounds—anything that lasts for a while, usually three years or more.

Durable goods come in two varieties: consumer and producer durables. Consumer durables include things like microwave ovens, gold bracelets, and automobiles. Producer durables refer to machinery and equipment. Durable goods are usually postponable purchases, so this kind of spending is a sensitive indicator of economic conditions.

Ee

EASY MONEY

Easy Money

A seemingly blissful situation created by the Federal Reserve Bank to reduce interest rates by expanding the amount of credit available. Blissful because the combination of lower interest rates and more credit means that borrowers who had been shut out of the debt market can now go in to borrow money—"cheaply." Seemingly blissful because the long-term result might be that money lent for dubious projects cannot be paid back when easy money becomes tight again. Which is what always happens. See TIGHT MONEY.

Econometrics

The use of mathematical and statistical techniques to solve economic problems, to test economic theories, and to predict the future. Econometrics is what gives economics its claim to being a "hard" science rather than a soft, social science.

Econometricians follow scientific methods. They formulate a mathematical model of how the real world—or a little bit of

the world—works. Then they use data from the real world to quantify the variables in the model and statistical techniques to check whether the quantified relationships are valid or just happenstance. If the model appears to be valid, it can be used to forecast the course of the world—or a little bit of the world.

Consider a simple—very simple—example. Say econometricians wish to test the hypothesis that the amount of consumer spending depends on the size of disposable income. They gather data on past amounts of consumer spending and the size of disposable income. From that, they infer the relationship between spending and income—or, what percent of income was spent over time. If it looks as if the relationship between the two variables is solid—say, that people routinely spend 75 percent of their disposable income—then econometricians claim that their model will forecast the dollar figure for consumer spending once disposable income is known.

Econometrics has all sorts of applications. To cite just two well-known ones: business uses it to forecast the demand for their products; government uses it to determine the impact policies will have on the economy.

There are, however, plenty of objections to econometrics. On a lofty plane, critics argue that the world, or any small part of it, is too complicated to be captured with mathematical precision. Or that, even if relationships in the real world can be quantified, relationships change too fast to be captured by models based on past experience. Less exalted—but no less reasonable—criticisms are based on the fact that many models fail to predict the future. That criticism holds especially true for the giant models of national economies which involve hundreds of equations which have to be solved together.

Nonetheless, the attempt to test theories with real-world data does set economics apart from other social sciences. (And as long as econometrics doesn't claim too much for its methods, knowing what doesn't work is probably just as valuable as knowing what does.) See MODEL.

Economic Efficiency

A condition that is achieved when resources are used without undue waste, cost, and effort.

The concept of efficiency is central to economics. Indeed, it is usually trotted out at the beginning of textbooks as part of the serious, even stern, discussion of how economics is a science that deals with the allocation of scarce resources. The point, of course, is that since resources are finite, they should be employed as wisely as possible.

Often textbooks then go on to draw an equally serious and stern distinction between efficiency and equity. This usually involves a lesson on how an efficient economic system isn't necessarily fair to all participants—an efficient economy might, for example, make some players richer than others. While the distinction might seem unnecessary, it isn't. Many economists use "efficiency" as a kind of stamp of approval. Since economic efficiency is what society is supposed to achieve, calling something efficient has come to mean good and branding something inefficient has become a code word for bad. See ALLOCATION OF RESOURCES.

Economic Growth

This happens when the productive capacity of the economy increases. Growth rates are usually measured by changes in gross national product with inflation subtracted. By this index, then, economic growth occurs when the real value of the goods and services generated by the economy goes up.

Economists are exceedingly fond of talking about economic growth in terms of pies: The more productive the economy, the larger the pie. Now, granted, most of us would choose a big pie over a small one, but crude size misses some important considerations.

Since the point of having a pie is to eat—and enjoy—it, the size of each person's slice becomes relevant. Here the variations are several: If population grows faster than the pie, each person will have a smaller slice to eat. Alternatively, if

ECONOMIC GROWTH

population shrinks while the pie remains constant, everyone gets a larger slice. In other words, a growing pie is not strictly associated with growing well-being. Hence, true aficionados measure economic growth in terms of slice size or as gross national product per capita.

Furthermore, growing pies and/or slices are no guarantee of deliciousness. If the pie is growing because everybody is working harder and enjoying less leisure, then the pie can be said to be less tasty. Or, if the pie and/or slices are growing because people are denying themselves the pleasures of current consumption then, again, growth comes at the expense of enjoyment. Again, a bigger and bigger economic pie doesn't always ensure well-being.

Economic Indicators
A bunch of statistics that are supposed to describe or forecast changes in the business cycle.

Indicators come in three clumps. The leading indicators (like new orders) are supposed to rise before economic upturns and fall before downturns; the coincident indicators

(like the unemployment rate) are supposed to rise and fall at the same time the economy does; and the lagging indicators (like consumer debt) are supposed to rise and fall after the economy changes.

Sounds real simple. And if it worked as neatly as it sounds, then forecasting economic activity would be child's play. Needless to say, however, the relationship between the indicators and the economy is not perfect. Although the indicators are right more times than they are wrong, there is no way of knowing for sure when it's safe to believe in them. Smart forecasters, thus, view the indicators with skepticism.

Economics

What economics textbook writers say you'll learn if you read their books, what economics professors say you'll know if you attend their classes, what economists say you'll profit by if you follow their advice.

Alternatively, economics is what noneconomists argue is a meaningless exercise, and what cliché mongers call the dismal science.

That having been said, what is economics? Most generally, it is a roughly scientific study of how the allocation of scarce resources do and do not lead to the satisfaction of human wants. Most specifically, it is what will not confound you if you use this dictionary as a shield.

Economies of Scale

This is what happens when a firm or industry expands its total production but reduces its costs per unit. In other words, as the scale of production gets larger, the enterprise becomes more efficient.

There are several major ways a firm can realize the benefits from economies of scale. First, it can take advantage of something called indivisibilities. Consider a business prop—like a

desk—that is not, strictly speaking, divisible. A firm buys and maintains a desk even though there might not be enough paperwork to keep an employee sitting at it all day. But when the business expands, the employee can use the desk from dawn till dusk without extra cost to the firm.

Second, a firm can realize economies of scale through specialization and the division of labor. A one-person carmaker, for instance, has to assemble the entire car. But when the firm expands production and hires more workers, each worker can concentrate on one task, thereby becoming more expert and more productive.

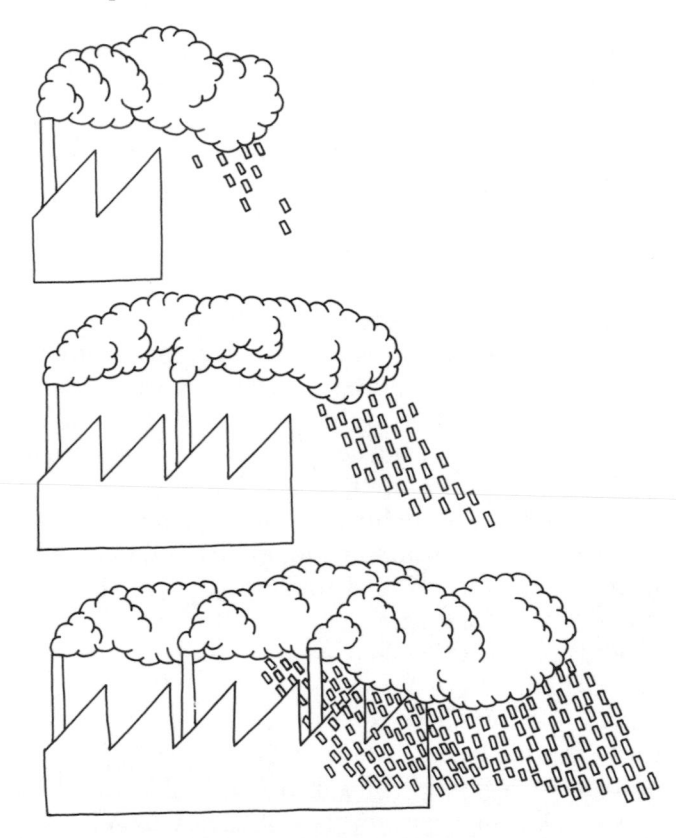

ECONOMIES OF SCALE

Third, a firm can gain from economies of scale by using larger, more efficient machinery. In the carmaking example above, a firm making only a few cars a year might simply hand-paint those cars. But when the firm expands to the point that cars are coming down the assembly line every minute, it can profitably use an automatic paint-spraying machine.

The existence of economies of scale is often taken as a reason why firms should grow bigger. And, sure, there are distinct advantages to bigness. In addition to the benefits outlined above, bigger firms—with lots of physical assets and cash reserves—find it easier to raise outside capital. Also, bigger firms are better suited to fostering research and development which often requires a pricey, long-term commitment. Bigger firms, too, can market their product more efficiently, perhaps by getting price discounts for buying in large volume or by arranging their own distribution networks.

But there are also arguments against the bigger-is-better theory. These, appropriately, fall under the category of diseconomies of scale. Here, bigness results not in greater efficiency but in inefficiency.

A classic diseconomy of scale in large corporations comes from the managerial side. Not only is it difficult to direct big operations, but bureaucracies themselves often shelter a lot of lazy—or downright incompetent—people. Too, it is thought that when workers are alienated from management, they become malcontent.

And, finally, some forms of business are just better smaller. Service firms that need to maintain good contact with their clients are better off on a smaller scale. Smallness might even make sense for some goods producers: a shop producing one-of-a-kind, super-luxe cowboy boots, for example. On balance, then, where does this leave us? With that most useful phrase, "it all depends." For some firms, getting bigger will mean getting better. For other firms, getting bigger will mean getting worse. And for still other firms, the better and the worse will cancel each other out. Every firm has its own optimum size.

Elasticity
A measurement of the responsiveness of one variable to changes in another.

The concept of elasticity is very important in economic analysis, and really not as abstract as the definition sounds. So, before you nod off, consider a specific example, what economists call the price elasticity of demand. This is designed to show how consumers respond to price changes. Now, common sense dictates that if the price of something goes up, consumers will buy less of it. The tough part is knowing whether consumers will cut back their purchases drastically or just a tad. And that is where elasticity comes in.

Elasticity is calculated by dividing the percentage change in the quantity that consumers demand by the percentage change in the price. The resulting number, called the elasticity coefficient, indicates the magnitude of consumer reaction. If the coefficient is less than 1, demand for that good is considered "inelastic," which means that consumers will cut back purchases only a touch in response to a price increase. If the coefficient is more than 1, demand for that good is considered "elastic," which means that consumers will cut purchases way, way back if prices go up.

As a rough rule, goods or services for which demand is inelastic include things that have no close substitutes—like salt—or that constitute a small part of consumer purchases—like paper clips. Goods or services for which demand is elastic include things that have close substitutes—like lamb chops—or that take up a large part of consumers' budgets—like cars. See DEMAND; SUPPLY.

Entrepreneur
In economic theory, the definition is lifeless—to wit, the owner-manager of a firm. In economic lore, however, the definition is larger than life itself—to wit, a risk-taking,

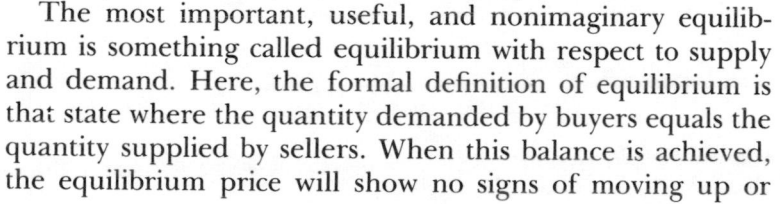

ENTREPRENEUR

clever-thinking, farseeing, establishment-defying head of his or her own enterprise, with the emphasis on enterprise.

Equilibrium

A state where the entire economic system is in balance. This state is largely mythical, but that certainly doesn't prevent economists from talking about it. However, to avoid the charge that they are indulging in magical thinking, economists say the economy "tends" toward equilibrium.

The most important, useful, and nonimaginary equilibrium is something called equilibrium with respect to supply and demand. Here, the formal definition of equilibrium is that state where the quantity demanded by buyers equals the quantity supplied by sellers. When this balance is achieved, the equilibrium price will show no signs of moving up or

EQUILIBRIUM

down. (Equilibrium price is also called the market clearing price.) See DISEQUILIBRIUM; SUPPLY AND DEMAND.

Exchange Rates. See FOREIGN EXCHANGE.

Expectations

Beliefs about the future. While that might sound harmlessly abstract, expectations have a real and important impact on economic decisions. Firms base pricing decisions on expectations about their economic prospects and the direction of costs; consumers base purchasing decisions on expectations about their future earnings and the direction of prices.

But having said that, it is hard to say much more with any certainty. Expectations are crucial but, since they cannot be directly observed, they are pretty much unmeasurable. Economists have tried to slip around the measurement problem by concentrating, instead, on how expectations are formed and why they change.

Export

Anything that's made at home but sold in another country. Exports can be tangible things like computers and corn, or they can be intangible ones like banking and insurance services.

Countries like to export, so governments usually have policies to encourage the activity. But because governments must pay lip service to the idea of free trade, they don't like to pay producers directly to export. Instead, governments do sneaky things, like subsidizing exports by offering producers low interest rate loans. See IMPORT.

Externality

A side effect, or spillover, from an activity that does not naturally have a market price.

Externalities can arise from either the consumption or the production process. For example, if your neighbor turns his front yard into a parking lot, the unsightliness that greets you when you look out your window is called a consumption externality. And if a businesss dumps its industrial waste in a river, the resulting pollution is called a production externality.

(Externalities aren't always bad. If your neighbor creates a formal English garden in his front yard, he gives you a rather pleasing consumption externality. Usually, however, people focus on the bad ones and take the good ones as their due.)

The slippery thing about externalities, at least for economists, is that there is no easy way to fix a price on them. Or enforce it. Just imagine the hassle in making your neighbor-with-the-parking-lot pay for the ruination of your view. (Likewise, your neighbor-with-the-English-garden would have a tough time making you pay for the improvement.)

Thus, the existence of externalities provides a good case for government intervention. The government, in effect, can put a price on externalities, either by taxing the activity that leads to them or by levying a fine on them. The government may also prohibit certain activities outright, like forbidding people from turning their yards into parking lots.

Ff

Factor of Production

A resource that is used to produce goods or services.

Factors of production are, arbitrarily, divided into three categories: land or natural resources, such as minerals, timber, water, and land itself; labor, such as physical and intellectual talents; and capital, such as factories, transportation, and machinery. (Sometimes economists include a fourth factor of production—the entrepreneurial skill that goes into organizing the first three factors.)

Each of these three factors is considered a scarce resource and is compensated for its contribution to production. For land, the compensation comes in the form of rent. For labor, payment comes in the form of wages or salaries. And for capital, income comes in the form of interest. (Entrepreneurial ability is rewarded by profit.)

All of these words are just definitions, of course. What's really interesting about the factors of production—the way they are organized, how they are priced—is the subject of economics. See CAPITAL; LABOR; LAND.

Federal Open Market Committee. See FEDERAL RESERVE SYSTEM.

Federal Reserve System

Refers to both the entire Federal Reserve System and its most visible part, the Federal Reserve Board. The Fed, to use its nickname, is our central bank, the bank of banks; its clients are commercial banks and the government.

The Fed was created in 1913. It consists of twelve banks spread across the country and a board of governors located in Washington. All the interesting action takes place in Washington, where seven governors and the chairperson direct functions that make the Fed a very heavy hitter in determining the course of the economy.

The Fed's most important function is to control the money supply. This is accomplished through two activities: buying

and selling government securities and changing reserve requirements. The key to understanding these activities is knowing how the Fed interacts with its member banks. This requires a rather lengthy explanation, but once this relationship is understood, everything else will fall into place.

Banks are required to keep reserves against the money they loan. If you open an account at a bank by depositing $1000, the bank cannot turn around and lend out the entire $1000; it must reserve some fraction of that amount. Banks that are members of the Federal Reserve—which includes all big banks and most small ones—are required to keep a minimum amount of reserves against their deposits. These reserves are usually held as deposits in the Federal Reserve System. It is through its grip on these bank reserves that the Fed is able to keep its hand on the loan activities of banks and, thus, on the economy.

Now back to just how the Fed controls the money supply. The Fed's most powerful activity, the buying and selling of government bonds, is carried on by something called the Federal Open Market Committee (FOMC) that does something called open market operations. Say the Fed thinks that the economy is overheating—business is expanding too fast and/or inflation is rising too steeply—then the Federal Open Market Committee will move to restrict the amount of credit. It will sell government securities.

The buyers of these securities pay the Fed with checks written on their banks. When the Fed presents these checks to the banks, they pay out of their reserves. The banks end up with less reserves and less money available to loan. (The banks have to cut back on the amount of loans they can make or their reserves will fall below the minimum requirement.) And when banks restrict loans (or credit) activity, the economy slows.

The second, but less used, device to control the amount of credit in the economy is the Fed's power to set reserve requirements. In this case, all it has to do is increase the reserves that it requires its member banks to keep on hand. Thus, with

a larger fraction of deposits going into reserve accounts, banks have less money to lend out.

A third, and also critical, function of the Fed is to act as the lender of last resort. This activity is accomplished through what's called the Fed's discount window. (The discount window is the fanciful name given to the Fed's own lending activities.) When its member banks are short of cash, they can borrow money from the Fed for a fee, known as the discount rate. When the Fed wants to encourage bank credit activity, it lowers the discount rate, making it cheaper for banks to borrow. When the Fed wants to restrict activity, it raises the discount rate.

Lending as a last resort is required when some event causes financial panic. Say that a big bank gets stuck with a lot of bad loans. Once the word gets out that the bank is having "difficulty," depositors start appearing to demand their money. If enough depositors remove their money, the bank will have to close its doors. (A fanciful way of saying that the bank has run out of ready cash with which to back, or refund, deposits.)

Of course, a bank with closed doors means that no depositor can cash in his or her account, and that people will stop accepting checks drawn on the closed bank. If the panic gets out of hand, depositors at banks that are still sound suddenly start demanding their money. These banks will have to use their reserves and, when their reserves run out, start calling in their loans, or shutting their own doors. In other words, the banking system will be faced with a first-class liquidity crisis.

But the Fed, by making cash available to the banks, can stop the panic. Simply put, when the Fed motors out cash to banks through its discount window, anxious depositors realize that the Fed stands ready to ensure that they can have their cash whenever they want it. Thus, depositors will be less likely to all show up at the same time demanding that money, and banks can go on about their business of taking deposits and making loans.

One of the more controversial functions of the Fed, especially in the past few years, is its role in financing government

budget deficits. When the government spends more than it collects in taxes, it covers the difference by borrowing from the public. That is, the Treasury issues debt securities in the form of notes, bills, and bonds. More colorfully put, the Treasury prints some money.

If the Fed decides to go along with the Treasury, it will buy those debt securities, thus injecting money into the financial system. How? By just the opposite process of what happens when the Fed sells government securities as described above. When the Fed pays for its purchases of Treasury securities, it increases the reserves held by its member banks. And with more reserves on hand, the banks suddenly have more money to lend. This whole process is called monetizing the debt.

As this country's central bank, the Fed performs a few other, less dramatic, functions. It keeps an eye on its member banks' lending practices and other day-to-day activities, and handles the country's foreign exchange operations.

Net-net, then, the Fed is an extremely powerful institution. It controls the nation's money supply and, thus, is an important player in determining whether credit is easy or tight, whether prices go up or down, and whether times are good or lousy. For instance, the Fed refused to act as the lender of last resort during the Great Depression and thereby prolonged it; it pumped out too much money in the 1970s, thereby contributing to double-digit inflation; and it clamped down sharply in the early 1980s, thereby causing a nasty recession.

In fact, some economists—and a lot of journalists—call the chairperson of the Fed the second most powerful person in the country. That, of course, is just the kind of oversimplification that gives both economics and journalism a bad name. See BANKS AND BANKING; OPEN MARKET OPERATIONS.

Final Goods and Services

Goods and services that are purchased for final use or consumption. Final goods are distinct from intermediate goods and

services which are inputs into the production process. The distinction is an important one for measuring the economy.

The mother measure for economic growth, the gross national product, is calculated by adding up the value of final goods and services; otherwise, the many intermediate products that go into making a final product would lead to double counting. Consider the production of the simple scoop of ice cream. The dairy farmer sells milk to the ice-cream maker for $1. The maker combines the milk with sugar and flavorings and sells the ice cream to the retailer for $3. The retailer then sells it to the final buyer, the consumer, for $5. GNP ignores the $4 of input and counts the value of the ice cream at $5. Thus, GNP can offer a reasonable estimate of the value of all output in the economy. See GROSS NATIONAL PRODUCT.

Financial Intermediary

An institution that links savers with borrowers. Financial intermediaries include banks, savings and loan associations, finance companies, insurance companies, and pension plans.

The process of financial intermediation is simple. Savers are those who wish to put their own money to work—that is, they want to earn money with their money. Borrowers are those who wish to put other people's money to work. Financial intermediaries stand between these two groups, as middlemen, and pay savers a fee for lending money and charge borrowers a fee for using it. See BANKS AND BANKING.

Fine Tuning

The conceit that the government can delicately intervene in the economy to promote growth and employment.

The idea of fine tuning was put forward by Keynesian economists and politicians during the 1950s and 1960s. The tuning itself was to be achieved through changes in the tax system and/or government spending. (Economy slumping a bit? Then inject a little spending oomph through lower taxes or more government expenditures.)

As it turned out, however, there were several problems in the actual execution. First, there is a lag between the time that something goes wrong and policymakers recognize it, mostly because the collection and interpretation of data on the economy take time. Then, there are further lags until the correct solution is hit upon and agreed to. After that, there are more delays until the solution kicks in, mostly because policy actions take time to filter through the economy. Finally, the economic data themselves can be misleading, so that even if there were no lags, policymakers might mistakenly tune a well-running machine.

At any rate, happy though the thought is, it doesn't work. So, most economists now agree that fine-tuning fiscal policy is an act of hubris. See FISCAL POLICY.

Firm

A business unit that produces goods or services. A firm hires labor, buys other goods and services, and then combines those factors of production to make something that it can sell. Firms hope to earn profits by their activities.

Consider a firm like a pizzeria. To produce pizza, the owner of the firm might rent a place of business, install various equipment like an oven (and a cash register), hire a pizza maker, buy the ingredients for dough, sauce (along with cheese, sausage, and whatever), and then organize all the above to produce pizza. The point, of course, is that the firm can then sell the output—pizza—for more money than it needs to pay for all the inputs.

Ownership of firms can range from a single individual to millions of stockholders. A group of firms in the same business constitutes an industry.

Fiscal Policy

Government policy that has to do with taxation and spending.

On the face of it, the workings of fiscal policy sound a little,

well, silly. First, the government takes money away from its citizens by taxing them. Then, the government turns around and spends that money on goods and services for its citizens. (And, of course, it charges a fee—salaries for bureaucrats—to move the money around the circle.)

For conservatives, making fiscal policy gives the government the right to act like a self-serving potentate, snatching a gold ring every time its money merry-go-round sails past. For liberals, making fiscal policy allows the government to play Robin Hood, taking money from the rich and spending it on the poor.

But, for some economists, making fiscal policy gives the government a grander role than just spinning money through the economy. By changing what's called "the mix" of taxes and expenditures, the government tries to achieve full employment and price stability.

If, for instance, policymakers fear that the economy is overheating—people are spending too feverishly and inflation is threatening to ignite—fiscal policy offers them two recourses.

First, the government can increase taxes. Raising taxes reduces the amount of money people have to spend. And when consumption falls off, so should prices. In fact some fiscal policies, like a progressive income tax, are thought to be automatic stabilizers: When economic activity is robust—and people and businesses are earning a lot of money—they automatically pay more in taxes. Second, the government can cut back on its own expenditures which would also dampen the demand for goods and services.

The trick, of course, is for policymakers to find the magic point where the economy slows down just enough to control inflation without creating higher unemployment.

The key to whether government is trying to stimulate or slow the economy is whether the government runs a surplus or a deficit in its budget. (In the above case, the government would be spending less than it takes in—the budget should show a surplus.)

The trouble with this fiscal policy, at least in recent years, is that it doesn't seem to work. Not only does government spending policy have one direction—up—but, despite the fact that the budget has been in seemingly permanent deficit, full employment has not been achieved. So, too, tools like the progressive income tax have not been effective in combating inflation. See FINE TUNING.

Fixed Cost

A more precise term for overhead. Fixed costs do not depend on a firm's output: for example, rent and interest payments on existing debt must be paid even if the firm isn't producing one single widget.

Fixed costs are only fixed in the short term, however. In the long term, of course, all costs are variable. A firm can always move offices or reduce its debt. See VARIABLE COST.

Flat-Rate Income Tax

A single rate of taxation for all taxpayers, regardless of the size of their income, and on all income, regardless of its source.

A flat-rate tax is appealing in several ways, mostly perhaps out of disenchantment with the current system. First, the current system's progressivity, a schedule of steeper and steeper marginal rates, dampens incentives to work harder—or longer. Bluntly put, if the more you earn, the less you keep —why bother? A flat tax, of course, is not progressive—the more you work, the more you earn, and the more you keep.

Second, the current system can be unfair. Rich people pay a larger share of their income in taxes than poor people. That inequity seems harmless, however, compared to the fact that the current system's maze of loopholes allows people with identical incomes to pay different amounts in taxes. The flat tax argues that not only should all people face the same tax rate, but that people with identical incomes should pay identical taxes.

Finally, and perhaps most appealing, a flat-rate tax with no exemptions, preferences, or credits would be a snap to figure. In fact, a true flat-rate income tax would be so simple that the entire form would fit on a postcard. See TAXES.

Foreign Exchange; Foreign Exchange Market; Exchange Rates

The first two terms are a snap to define. Foreign exchange refers to claims to hard cash in the currencies of foreign countries. The foreign exchange market is where foreign exchange is traded; it's an international marketplace whose players, connected by telephones and computers, arrange for the transfer of currencies from one country to another.

But the third term, exchange rates, is not so easy. Quickly defined, an exchange rate is the price at which one currency can be traded for another. But that definition raises two questions: Why are exchange rates necessary, and how are they determined?

Exchange rates wouldn't be necessary if countries did not engage in foreign trade. Trade is an uneven process. Countries have different fiscal and monetary policies, so they have different growth rates, interest rates, and inflationary expectations. Countries also have different cost structures, so that prices of identical goods may vary from country to country. Or countries may make unique products. All these differences mean that some countries have so-called strong currencies—ones in great demand—while others have weak currencies—ones for which demand is slight. Thus, trading partners need a mechanism that converts the value of one currency into the currency of another.

Consider what happens when Americans buy French wine. Americans wish to trade their dollars for wine, but the French wish to trade their wine for francs. So dollars must be exchanged for francs if the trade is to go forward.

The exchange rate indicates how many francs—and thus, how much wine—a dollar will buy. Hence, the exchange rate

between dollars and francs gives the American buyer and the French seller the chance to evaluate the trade in terms of their own currencies. If the exchange rate is 9 francs for $1, then $100 will purchase 900 francs worth of wine in France.

And now for the second question: How are exchange rates determined? Here, the answer depends on what kind of exchange rate system countries choose to operate under. There are, basically, three choices: a fixed rate, a floating rate, or a managed rate.

Under a fixed rate system, countries agree to a set ("par") value for their currencies. For example, the exchange rate between German marks and American dollars could be set at 3 marks to $1, or the exchange rate between French francs and American dollars could be set at 9 to 1.

The central banks of each country must then agree to hold reserves of all currencies and to buy and sell from these reserves to offset deviations from par values. If, for example, an excellent vintage of French wine caused a rush of wine buyers in other countries to demand francs, the value of the franc would rise above its par value. In that case, the French central bank must sell francs and buy other currencies in order to restore equilibrium.

Under a floating rate system, on the other hand, the market sets the value of currencies based on supply and demand. Say the demand for dollars is great—either because of high interest rates, good investment possibilities, low inflationary expectations, or fierce desire to buy American goods and services—then the dollar's exchange rate price will be high. A floating rate system means that exchange rates are flexible, free to change from hour to hour, or minute to minute.

Needless to say, a floating rate regime means that central banks cannot intervene in the market to adjust the price of currencies. For example, if a great vintage causes heavy demand for francs, the central bank of France would just watch as the franc appreciated against other currencies.

Under a managed float, the third type of exchange rate regime, countries say they are on a floating rate system but

act as if they were under a fixed rate one. That is, central banks allow the market to set the exchange rate—unless they don't agree with the result. The net of a managed float, also called a dirty float, is that currencies appreciate and depreciate only as much as central banks permit.

During the 1950s and 1960s, most major currencies were on a fixed, but adjustable, regime. Currently, most major currencies are on a floating, but managed, system. Neither of those regimes seem to work very well, so the debate over how an international monetary system should be run is still a hot one. And an important one.

There are several arguments against floating exchange rates. Most dramatically, when the foreign exchange price of a currency goes up and down, trading relationships are disturbed and the domestic economy suffers. Say the dollar depreciates against the currencies of our trading partners. While that makes our exports cheap, it also makes foreign imports expensive. Eventually, as demand for foreign imports shrinks, business and jobs in the import sector will be lost.

Too, floating rates can be inflationary. In terms of a depreciating dollar, for instance, the rise in the dollar price of imports both provides a shield for domestic industries to raise their prices and pushes up costs for industries dependent on imported raw materials.

Furthermore, without the discipline fixed rates impose on central banks to keep their currencies at an agreed-upon value, central banks can—and do—pursue inflationary increases in the money supply. (Under a fixed rate system, if banks create too much money, their currency starts to depreciate and they are then bound to start buying their own currency—thereby reducing the supply—in order to restore exchange rate equilibrium.)

There are also, of course, several arguments against fixed exchange rates. To begin with, the requirement that governments keep their currencies stable can create distortions in trade. A country faced with an appreciating currency might not wish to undergo the often painful measures needed to

bring its currency back in line. Instead, the government might try to restrict imports by passing protectionist laws.

So, too, central banks can always weasel around the discipline of the fixed rate system. Say a dollar depreciation caused the Fed to go out and buy up dollars abroad; the Fed can then turn around and put those dollars back into domestic circulation by buying bonds. See APPRECIATION; BALANCE OF PAYMENTS; DEPRECIATION; GOLD STANDARD.

Foreign Exchange Market. See FOREIGN EXCHANGE.

Free Trade
A situation in which governments do not interfere with the international exchange of goods.

Free trade is an extremely desirable event. It encourages competition, and thus the efficient allocation of the world's resources and economic growth. Nonetheless, most governments persist in policies that inhibit it. They affirm their commitment to free trade, but fight to retain export subsidies, import quotas, and tariffs. See PROTECTIONISM.

FREE TRADE

Frictional Unemployment

Joblessness that is temporary.

Frictional unemployment is the result of imperfections in the labor market. When people decide to leave jobs, or when their jobs disappear, it takes time to find other employment. And, during that gap, they are counted as unemployed. Frictional unemployment can't be eliminated, but it can be reduced by better information and increased job mobility. See UNEMPLOYMENT.

Friedman, Milton (1912–)

Father of monetarism and Nobel prize–winner. Friedman is a spectacular powerhouse: He has had two successful careers —as an eminent academic economist and as a formidable propagandist.

As a longtime professor of economics at the University of Chicago, Friedman created a theoretical counterpoint to Keynesian economics known as monetarism (or, more informally, as the Chicago School). Friedman and his disciples argue that the amount of money in the economy is the most important determinant of prices, and that major disruptions in the economy come from politicians fooling around with the money supply. Much of the evidence to support this argument comes from his 1963 book, *A Monetary History of the United States 1867–1960*, written with Anna Schwartz.

Friedman's political views flow quite smoothly from his academic work. Since he believes that most economic problems are caused by meddling politicians, especially those sitting on the Federal Reserve Board, his remedy is to have a nonactivist government, to free the markets from laws and regulations designed to direct economic activity. Friedman has argued his views tirelessly in books, magazines, and newspapers, on television, and in person to presidents, congressional committees, and heads of foreign states. See MONETARISM.

Full Employment

When most everybody who wants to work, does.

Note that full employment does not mean that everybody has a job. Total employment is just not possible. There will always be some workers who are temporarily unemployed, either because they are in transit from one job to another or because their work is seasonal. The figure for full employment varies; some economists put it now at 94 percent (a 6 percent unemployment rate).

While full employment is a perpetual goal of governments, it is rarely achieved.

Gg

Galbraith, John Kenneth (1908–)

Economist, liberal politician, novelist, essayist, but mostly gadfly.

Galbraith is unlike most well-known economists in that his chief contribution has nothing to do with scholarly work, but with his popular writings. And unlike most economists, well-known or not, Galbraith is witty, worldly, stylish. Indeed, Galbraith is not an economist in the modern sense: He rails against what he calls conventional economic wisdom with elegant prose, not with statistics and mathematical models.

He argues, for example, in favor of an interventionist government, of the superiority of public goods over private consumption, and he asserts that a "technostructure" (a managerial-professional elite) determines consumer tastes. A lot of these polemics found their way onto the best-seller lists in books like *The Affluent Society* and *American Capitalism*. Many of Galbraith's ideas are not original, and his arguments, or propositions, are not testable. Yet, even his critics would grant that, on the sheer force of his personality, he's an economist to be reckoned with.

Game Theory

Theoretical analysis of the decision-making process taken by two or more players who are in conflict.

Game theory looks at the possible strategies of players who have to make decisions without knowledge of what other players are planning. Each player's strategy, once undertaken, will affect the others. Take, for example, one of the best known parts of game theory—the Prisoners' Dilemma. Here the players are all prisoners, jointly accused of a crime, and held in separate cells. If nobody confesses, they will get medium sentences for lack of evidence. If one prisoner squeals, she will get a light sentence while the others will get hammered. If all confess, they will all get longer sentences than if they had all remained silent, but shorter than if just one had confessed.

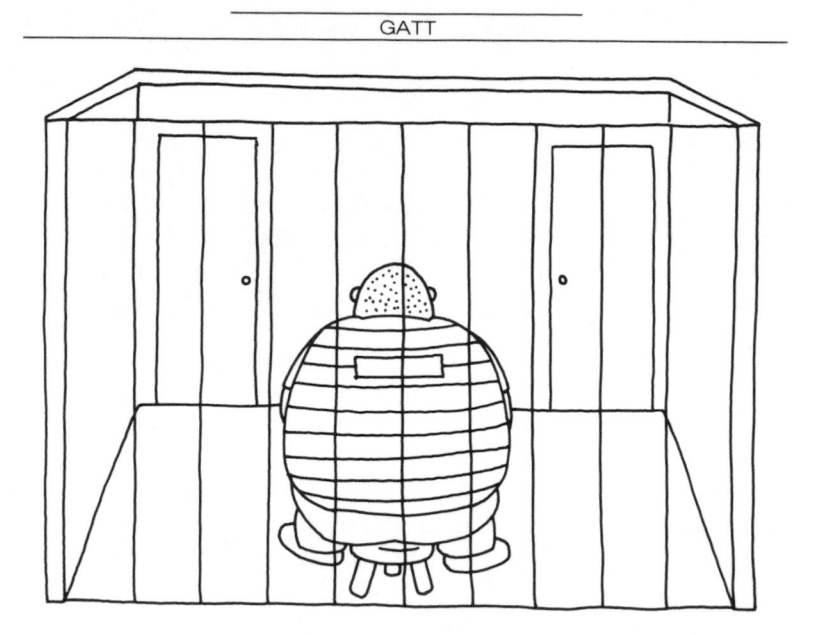

GAME THEORY

Obviously, each prisoner has an incentive to squeal, hoping to be the only one to do so and thus getting off with the least punishment. If all the prisoners squealed, however, they would all, so to speak, hang together. Only if all the prisoners act altruistically, by remaining silent, will they all benefit.

Game theory was developed by mathematicians in the 1950s—notably by John von Neumann. Economists started moving in during the 1960s and now game theory is used to explain things like the behavior of oligopolists. See ZERO-SUM GAME.

GATT

General Agreement on Tariffs and Trade. International rules of good behavior on trading matters.

The purpose of GATT is to promote trading among the world's nations. Thus, GATT tries to hold its members to a

code that forbids various antitrading practices, like certain types of tariffs and import quotas. These restrictions have been agreed to in periodic negotiations. Since GATT started operating in 1948, the general level of tariffs has, in fact, declined markedly. See PROTECTIONISM; QUOTA; TARIFFS.

General Equilibrium

A state in which all markets in the economy are stable; that is, when supply equals demand everywhere so there is no tendency for prices to jump about.

The analysis of general equilibrium is one of those grand concepts that are useful for understanding how the economy works in aggregate. Say a run of bad weather reduces the corn crop. The immediate result will be felt in the price of corn—less supply means higher prices. Simple. But general equilibrium doesn't stop there. Other prices, in other markets, are affected by the new, higher price for corn. The prices of products made from corn—corn flakes and corn muffins, for example—will rise also. But the price of labor in these markets might drop as higher product prices cause consumers to demand less and thus lead to a slowing of business. Too, the prices of products that are good substitutes for corn, like other grains, might rise as consumers demand less corn but more of the substitute products.

Quickly said, general equilibrium analysis assumes that markets are interrelated. A disturbance in one market will create disturbances in others, and the economy cannot be in equilibrium unless all its markets are.

The idea of general equilibrium was developed by French economist Léon Walras in the late nineteenth century. No doubt the idea is a powerful one; nonetheless, most economies are too complex and dynamic to capture, in practice, their workings in this manner.

GNP. See GROSS NATIONAL PRODUCT.

Gold Standard

A type of fixed exchange rate in which central banks are required to trade a fixed weight of gold for their currencies. In effect, gold becomes a sort of world currency: International debts are settled in gold and, since currencies are convertible into gold, differences among them disappear. But a gold standard does more than function as the world's money —it automatically links changes in a country's money stock with its balance of payments.

Say, for example, that a country exports more than it imports, thus running a surplus in its balance of payments. That surplus means the country is experiencing an inflow of gold, so the central bank can expand the money supply without worrying about having enough gold to cover those paper liabilities.

But consider what happens if the bank prints up a bit too much money: An increase in the money supply will push prices up, and higher prices will make it difficult for the country to keep up its strong exports. As the demand for its exports diminishes, so will its balance of payments' surplus. With less gold to back its currency, the central bank will be forced to cut back on money creation. Prices will fall and equilibrium will be restored. (The reverse is also true, of course. A country with a deficit in its balance of payments will be forced to reduce its currency to correspond to its gold reserves. As prices fall, its exports will become more attractive and it will experience an inflow of gold, thus reducing its balance of payments' deficit.)

A gold standard, then, controls inflation by disciplining the money creation of central banks, and, by creating stable prices through a fixed exchange rate, it encourages international trade.

If a gold standard can deliver such a tidy package, then why doesn't the world go back on one? Well, perhaps the most powerful reason is that many players feel that a country's balance of payments shouldn't control its money supply.

Often a government will wish to run a domestic policy that promotes employment and ignores inflation.

War, which requires a lot of money, is a perfect example of how countries can refuse to follow the discipline of a gold standard: The United States went off the gold standard during the Civil War, and Great Britain abandoned it during the Napoleonic wars and World War I.

The last serious attempt to hold the major trading nations to a gold standard came in 1944 with the system set up at Bretton Woods. Here, the dollar was on a gold standard and exchange rates for other currencies were fixed at a par value with the dollar. Whether the Bretton Woods system worked or not depends on whom you listen to; regardless, it was kiboshed in 1971 when President Richard Nixon put an end to the convertibility of the dollar into gold. See BALANCE OF PAYMENTS; BRETTON WOODS; FOREIGN EXCHANGE.

Gross National Product (GNP)

The mother measure of national economic well-being. The actual statistic is supposed to represent the dollar value of all goods and services produced by the nation's economy. It covers consumer expenditures for goods and services, private investment, and government spending and investment (but not transfer payments like Social Security).

The point of the exercise is to get a fix on whether the economy is expanding or contracting—and at what speed—on a year-to-year basis. GNP is so important, however, that the government issues estimates every three months; indeed, these quarterly looks are preceded by a "flash" or early estimate. So, too, the erratic course of inflation has made it necessary for the government to distinguish between GNP measured at current market value and GNP measured in constant, or deflated, dollars.

As with any umbrella statistic, there are megaproblems in calculating GNP. For one, the government must take care to count only goods and services for final consumption, not in-

termediate input: for example, the cheese on a cheeseburger should be counted when the cheeseburger is made, not when the cheese is produced. For another, although GNP is supposed to be all-encompassing, it ignores some goods and services: for example, the goods and services generated by unpaid housework, illegal activities, and the underground economy (where transactions go unreported to the taxing authorities) are not tallied.

Hh

EL BANCO

Hayek, Friedrich August von (1899–)

Winner of the Nobel prize in Economics in 1974 and distinguished member of the Austrian School.

Hayek has taught at half a dozen of the world's outstanding universities, and written more than as many books. Perhaps his most important work—and one that made international best-seller lists—was *The Road to Serfdom* (1944).

In this short but powerful book, Hayek laid out the dangers inherent in central planning and big government at exactly the moment most democracies were taking their first steps down this road. Hayek warned that economic planning cannot replace the free market, arguing that the journey toward a welfare state would lead to loss of both economic prosperity and political freedom. See AUSTRIAN SCHOOL.

Hot Money

Term applied to the large, and lightning-fast, shifts of capital among the world's currencies. (Actually "flighty money" is a more accurate description.)

Owners of hot money—often and unfairly described as speculators—sell a country's currency when they think it will depreciate, either because the country is on the verge of war, faces an economic slowdown, is experiencing an acceleration in the rate of inflation or just plain offers inferior investment returns. They buy a country's currency when they think it will appreciate, either because the country is politically stable (a safe haven), is about to enjoy a burst of economic growth, or offers superior returns, like high interest rates.

For central bankers and other politicians who are often burned by hot-money flows out of their currencies, the term is used disapprovingly. After all, hot money exhibits no national allegiance and its rushing around can increase the volatility of the foreign exchange markets. Nonetheless, there is absolutely nothing wrong with the owners of money seeking

HOT MONEY

high returns or safe havens—even if it makes them seem flighty.

Housing Starts

As it sounds—when ground is broken for new homes. Housing starts, which are estimated monthly by government bureaucrats from the number of building permits issued, are an important indicator of the level of economic activity.

Human Capital

A marketable commodity possessed by people in the form of education, skills, or talent.

Broadly put, human capital enables people to earn a living; specifically, it determines just how good a living they can earn. In fact, earned income can be considered a return on an investment in human capital. In this sense, doctors' fees are the return for their investment in a medical education.

Human capital has some of the same characteristics as physical capital. Indeed, a Ph.D. in economics is rather like a factory. Consider: Both assets require an initial investment—the owner must pay money to acquire them. Both assets involve some lag before the investment contributes to production. Both assets earn income over a period of time. And both, eventually, wear out.

Perhaps the most important shared characteristic between human and physical capital is that increasing a country's level of both kinds of capital is very important for economic growth—both raise productivity. (While one might question the assumption that a person with a Ph.D. in economics is more productive than one with a high-school diploma, few would argue with the notion that a literate person is more productive than an illiterate one.)

Governments routinely provide special incentives for citizens to improve their human capital by offering free schooling, providing student loans, establishing vocational schools, or supporting job training programs.

Of course, people don't undertake to improve their human capital with an eye to raising the general level of competence in a society. Usually they invest in themselves because the payoff in increased income can be a handsome one. (Not all investments in human capital are motivated by the economic reward, however. It's certainly true that the pursuit of more education or new skills can be rewarding in a nonpecuniary way, as well.)

Hume, David (1711–1776)

Scottish philosopher and historian who also made important contributions to economic thought, particularly to international trade theories.

In Hume's time, most major trading nations were on the gold standard. And most tried to keep to mercantilist policies that, among other things, discouraged imports in order to avoid outflows of gold. (Think of importing as a process whereby citizens exchange their gold for foreign goods.)

Hume, however, argued that the balance of payments mechanism automatically worked to prevent permanent trade deficits (hence, gold shortages). His theory was simple: When a country ran a trade deficit, the resulting outflow of gold would reduce its money supply, causing prices to fall. At the same time, the country experiencing an inflow of gold would find its money supply increasing and its prices rising.

Two things would happen as a result. The country with the new, lower prices would start exporting more as other countries rushed to take advantage of bargain prices, and importing less as its citizens found prices abroad too high. Likewise, the country with new, higher prices would see its exports fall and its imports rise.

This situation would quite naturally bring more gold into the exporting country while draining gold from the importing country. In other words, Hume argued that the rise and fall of prices would automatically bring trade flows into balance so that no country would run a trade deficit (or surplus) for long. See GOLD STANDARD.

Ii

IMF. See INTERNATIONAL MONETARY FUND.

Import

Anything that's made in another country but bought at home. Imports can be tangible things like autos and bananas, or they can be intangible ones like banking and insurance services.

There are certain things that countries like to import—anything, for example, that cannot be made at home. Generally, however, countries do not like to import things that are made at home, even though imports might be cheaper or better quality, because the imports will displace domestic goods. Thus, countries often try to control the volume of imports by levying tariffs or quotas. See QUOTA; TARIFFS.

Indexation

Method of changing prices for goods and services by linking them to changes in the rate of inflation.

Economies with wickedly high rates of inflation often turn to indexing to protect people when prices for some goods and services increase more rapidly than others. Specifically at risk are people with negotiated wage contracts or holders of long-term bonds, who must watch helplessly as the real value of their salaries (in terms of the amount of goods and service they can buy), or their bonds, is eroded by fast inflation.

In the United States, most labor contracts contain some sort of index clause, typically one which requires employers to raise wages a certain amount for each bump up in the consumer price index. Ditto for Social Security payments—they're indexed too. And, in 1985, the United States began another form of indexation: income taxes. Here the point is to protect taxpayers from the evil interaction between inflation and the progressive tax system—called bracket creep.

Before indexation, when wages rose to keep pace with inflation, taxpayers found that their new, higher wages subjected them to a new, higher tax-rate bracket, even though their wages hadn't gone up in real terms. Indexation now adjusts brackets for the impact of inflation.

While indexation sounds like a commonsense approach to the problems caused by inflation, it has its critics. They argue that indexation will make people insensitive to inflation (no pain, no anxiety) and thus less willing to fight it. And certainly in countries where almost total indexation is the rule, inflation can skyrocket. (But supporters of indexation reply that even in countries without indexation, inflation can get out of control.) See CONSUMER PRICE INDEX; INFLATION.

Inelastic. See ELASTICITY.

Inferior Good

Anything that violates the "law" of demand which states that when the price of something goes down, more of it will be demanded; with an inferior good, when the price goes down, less is demanded.

Although some people might consider the existence of inferior goods an interesting paradox, the explanation is quite straightforward. Take textbook writers' favorite example of an inferior good—potatoes. And assume consumers thought potatoes were less desirable then, say, pasta. If the price of potatoes drops, two things could happen. Consumers might obey the law of demand and buy more potatoes, which have become cheaper relative to pasta. Or consumers, now feeling richer (they can buy more potatoes for each dollar) and preferring pasta, might buy more pasta instead of potatoes. Potatoes, then, are an inferior good in the sense that the richer consumers are, the fewer potatoes they'll buy.

Inflation

Generally, rising prices. Sounds simple enough. But the question of what causes prices to rise has several answers.

Two explanations are known as demand-pull because prices are said to be pulled up by a sudden increase in demand, or spending. Monetarists blame demand-pull on monetary policy. Their argument runs like this: When the Federal Reserve pumps out more money than is needed for transactions, the new money will be used to purchase assets like stocks and bonds, television sets, and cars. Given that the amount of assets in the economy stays constant, this new surge in demand will bid up prices. (More money chasing the same amount of goods causes higher prices.)

A second demand-pull theory, put forward by Keynesians, agrees that a burst of new money will be spent on assets. But they see the primary impact coming from the purchase of bonds: Given that the amount of bonds remains constant, an increase in demand for bonds will bid up their prices which, of course, will lower interest rates. Lower interest rates, in turn, will encourage more business spending and some consumer spending. Again, the surge in demand for goods will pull prices up.

Another group of explanations are known as cost-push because prices are said to be pushed up by a sudden increase in production costs. The wage-push view blames labor unions which negotiate wages higher than can be justified by workers' productivity; companies then have to raise the prices of their products to cover these higher labor costs. The profit-push view argues that companies simply raise their prices to increase their profits. And the commodity-push view says that an increase in commodity prices—due to political decisions by major producers (like OPEC) or bad weather (like a drought in the grain belt)—will spread through the rest of the economy, pushing up prices in general.

Which theory is correct? Probably all of them. That is, there is no single answer. Inflation can be caused by a num-

ber of things, some more important than others at any given moment.

The causes of inflation may be muddled, but its consequences are not. Inflation can be bad for some people and good for others. For example, inflation makes losers out of creditors who are paid back in money that is worth less (that is, rising prices mean the money buys less) than when they lent it out.

On the other hand, inflation makes winners out of debtors who can repay their debts in cheaper money. (Some groups are immune from any impact. Wages generally keep up with rates of inflation—most union contracts provide for cost-of-living indexing. And Social Security and federal pensions are indexed, as are most private pensions.)

There is one kind of inflation, however, that is so disruptive to economic arrangements that it makes for losers across the board—hyperinflation. This is a condition of especially rapid and large increases in prices.

The classic example is Germany in 1922–23 when the average rate of inflation was 322 percent per month. Per month! Cash lost its value so rapidly that people became desperate to exchange it for goods. Workers demanded their wages twice a day so they could shop midday because, by the end of the day, prices would be higher. And—or so the joke goes—beer drinkers ordered two mugs at a time because the second beer would go flat slower than its price would rise.
See INDEXATION; KEYNES, JOHN MAYNARD; MONETARISM.

Innovation
Said to occur when a new and significant product or technique is introduced.

For economists, innovation is distinct from invention—and more honorable. Invention refers to mere discovery, while innovation refers to the development, promotion, or application of an invention.

INNOVATION

The term itself was popularized by Joseph Schumpeter, who recognized the crucial role innovation plays in economic growth. Imagine where we'd be without innovations like railroads, airplanes, and automation. (Or frozen foods and credit cards, for that matter.) Economists figure that innovation has accounted for half of the increase in our per capita productivity and real wages.

Of course, innovators—those who take the risks involved in applying inventions—are not motivated by grand visions for spurring economic growth. Rather, they are motivated by the profits that accrue to successful innovation.

Interest

The price paid over time by borrowers for the use of lenders' money. Or, more briefly, interest is the cost of borrowing money.

Interest is usually expressed as a rate, or percentage—I will lend you $100 (the principal) if you pay me 10 percent

(interest). Unless I am a loan shark, however, I cannot set the interest payment myself. Rates are usually set by the market; that is, rates depend on the supply of loanable funds and the demand for those funds. If, for instance, everybody is spending like mad and saving very little, interest rates will tend to be higher.

Economists distinguish between nominal and real interest rates. Nominal rates refer to the stated rates, like the 10 percent above. Real rates refer to the interest rate after the rate of inflation (or deflation) is taken out. Again, in the example above, if inflation is running at 5 percent, then the real rate of return on the $100 is only 5 percent.

International Monetary Fund

The IMF was established in 1944 as part of the Bretton Woods agreement. It functions as a sort of nanny for the international monetary system, chiefly by offering credit to nations experiencing what's formally called "balance of payment difficulties" and what's informally known as "bust."

Currently, the IMF has over 140 member nations. The fund itself is made up of money paid in by its members, mostly in their own currencies. The specific amount of these dues, known as quotas, depends on the relative size of each country's economy. In return for their contributions, the members get to vote on IMF policy and to borrow money from the fund; the bigger a country's quota, the larger its voting and borrowing power. The United States is the most powerful member.

In 1967, the IMF created something called special drawing rights (optimistically known as paper gold). The value of SDRs, which are issued by the IMF, depends on the value of the five national currencies that stand behind them. SDRs were meant to function as a sort of international currency, but so far their success has been rather limited.

Bankers like the IMF because it can—and will—bail them out when their loans go sour. Economists like the IMF be-

cause it issues a lot of good reports on international economic conditions. Liberals like the IMF because it adds an international air to helping needy countries. Conservatives, however, don't like the IMF because it is just one more intrusive bureaucracy. See BRETTON WOODS.

Inventory
The various kinds of goods that businesses keep around to ensure a steady stream of supply to their markets. Inventories include both finished goods and the raw materials needed for the production of finished goods.

It's good sense to keep inventories on hand. (Who doesn't try to maintain inventories of household staples like toothpaste and tomato soup?) But, for business, inventories can be tricky.

For starters, inventories are a form of investment, just like the construction of factories or the purchase of new equipment. And, since investment means a lag between the time that expenditures are made to obtain inventories and the revenue generated by sales, businesses try to keep inventories, thus expenses, low. That is easier said than done, however. While most firms try to maintain enough inventories to cover sales, plus some margin, few firms can accurately forecast sales. No surprise, then, that inventory-to-sales ratios can be variable.

The variability in inventory-sales ratios is more than an annoyance, however. It can have a major impact on economic activity. Say that sales slump and the ratio jumps up; in order to bring it back in line, firms will start cutting back on production. If the cutback is severe, or widespread, economic activity will slow and a recession could result. In fact, inventory reductions have been a major mover in post–World War II recessions.

Put another way, economic forecasters keep a narrowed eye on inventory-sales ratios published monthly by the government. Just as a rise in the ratio can be an early warning

sign that a slowdown is starting, a fall can indicate a pickup in economic activity.

Investment

Strictly—and properly—defined, investment refers to any expenditure where benefits accrue in the future.

Investment usually includes the construction of new office buildings, factories, houses, and apartment buildings, new equipment and increases in inventories. The term net investment means gross investment after allowances for depreciation. Economic investment applies to expenditures on capital goods while financial investment refers to expenditures for stocks or bonds.

Invisible Hand

Concept put forward in 1776 by Adam Smith in his *Wealth of Nations.*

The doctrine itself is simple: When each individual is free to pursue his or her own self-interest, the entire society benefits. That is, each is "led by an invisible hand to promote an end which was no part of his intention." What could be more appealing than doing good for others by doing good for yourself?

The "invisible hand" doctrine is more than a rationalization for selfishness, of course. It is a powerful argument in favor of free markets and competition. (And as such, it's quite attractive to economists and rather unattractive to politicians.) See SMITH, ADAM.

Jj

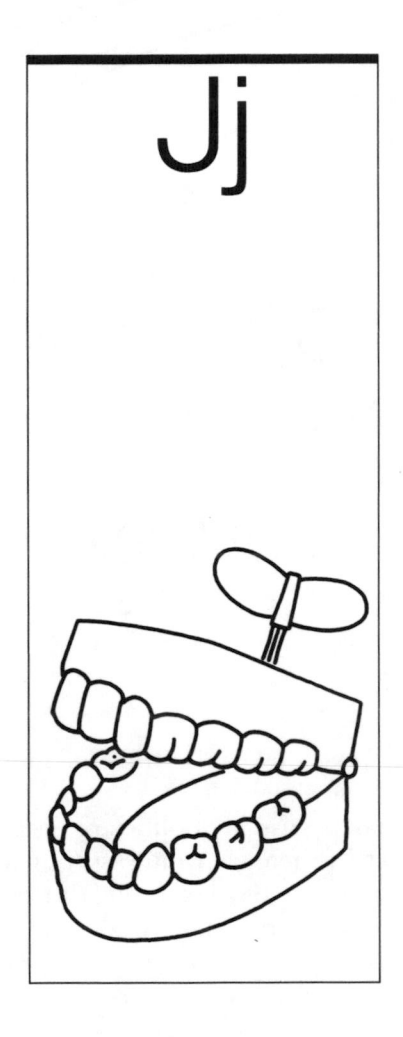

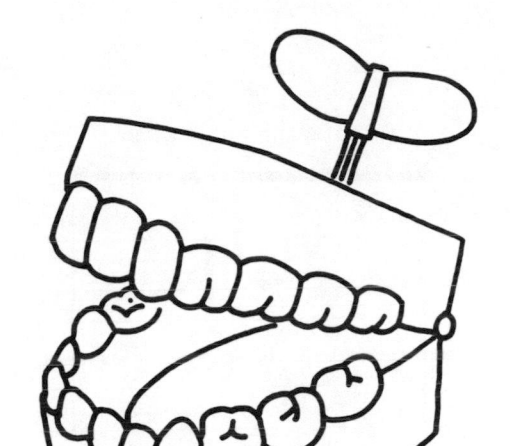

JAWBONING

Jawboning
When economic policy takes the form of moral exhortation or persuasion.

Jawboners, who are usually politicians, ask special interest groups not to act in their own interest, but to respond to a greater social imperative. For example, during periods of inflation, politicians jawbone labor unions to hold their wage demands down, or businesses to keep their prices steady. The conceit is that jawboning will embarrass these "naughty" groups into complying with a higher social good as seen by, of course, politicians. The nice thing about jawboning is that it is informal—no well-intentioned but silly laws are passed; the not-so-nice thing is that it is rarely successful.

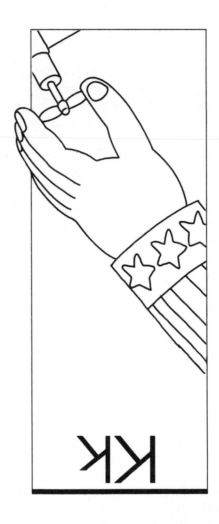

Keynes, John Maynard (1883–1946)

British economist, teacher, editor, government consultant, investor, literary groupie, Keynes was one of those fortunates who are superbly competent in whatever they turn their attention to. But it was as an economist that Keynes was truly awesome.

Keynes was the most influential economist of his generation. And of the following generation. But having noted that, I can't resist quoting Keynes himself on economists whose influence extends beyond their immediate time. He said: "Practical men, who believe themselves to be quite exempt from any intellectual influences, are usually the slaves of some defunct economist. Madmen in authority, who hear voices in the air, are distilling their frenzy from some academic scribbler of a few years back. . . ." In Keynes's case, the ideas he set forth in his *General Theory of Employment, Interest and Money* (1936) influenced the madmen in authority during the sixties and seventies.

Roughly, Keynes's insight was that the market was not self-correcting, at least not in the short run. The Adam Smith notion, which had dominated economic policy before Keynes, held that downturns would eventually become upturns. For example, when businesses found they couldn't sell their products, they would lower prices; as prices fell, demand for goods would pick up and the system would return to equilibrium.

The Keynesian notion, however, was that downturns could become permanent. That is, if people stopped buying, businesses would have to cut back production and lay off workers. Since these unemployed would not have money to spend, the economy might stagnate at low levels of production and employment. So, said Keynes, the government should step into downturns and, through its policies, goose up spending and get the economy moving again. (This notion was transformed slightly by the economists who followed Keynes and are called Keynesians: They argued that the government could fine-tune the economy and avoid downturns altogether.)

Specifically, Keynes proposed to pep up aggregate demand through government spending, even if it meant that government would have to run a budget deficit. With government spending substituting for the desirable level of private spending, businesses would hire more workers to produce more goods. What made Keynes's idea so revolutionary was the emphasis on government action: The government takes responsibility for maintaining employment.

The hooker, however, was the price level. While Keynesian theory assumes that prices remain constant, Keynesian practice proved different. During the late sixties, as the government spent itself silly to keep unemployment down, prices rose. And in the seventies, when government spending really took off, both prices and unemployment went up. Thus, in the eighties, Keynesians are in retreat and other theories about how the economy works—and doesn't work—have become popular. See FINE TUNING; MONETARISM; NEW CLASSICAL ECONOMICS; RATIONAL EXPECTATIONS.

Klein, Lawrence R. (1920–)

Klein won the Nobel prize in Economics in 1980 for his work in building econometric models. In the 1950s, he first constructed a giant econometric model of the United States. In the 1970s—no rest for the weary—he developed LINK, a technique that integrates econometric models of the world's major nations. See ECONOMETRICS.

Kondratieff Cycle

Named after the Russian economist, Nikolai Kondratieff, who argued that capitalist economies experienced long—fifty- to sixty-year—ups and downs. Kondratieff wrote in the 1920s. Joseph Schumpeter later adopted the so-called long-wave explanation which, he argued, was correlated with waves of innovation. Kondratieff cycles are different from Kuznets cycles. See INNOVATION; KUZNETS, SIMON.

Kuznets, Simon (1901–1985)

Winner of the Nobel prize in Economics in 1971, Kuznets pioneered work on economic growth, population growth, and national income statistics: but he will be remembered for his discovery of fifteen- to twenty-five-year swings in economic growth called, appropriately, Kuznets cycles.

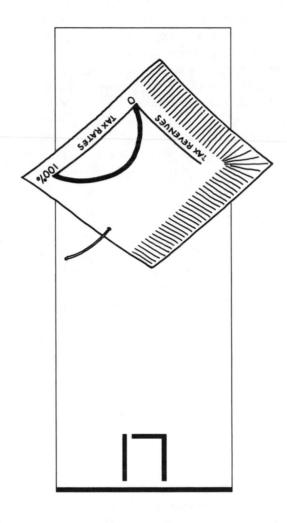

Labor

One of the three factors of production, land and capital being the other two. Labor input refers to any human effort—physical skill, intellectual power, or artistic skill—used in production.

Labor Theory of Value

The idea that goods and services should be valued solely by the amount of labor that goes into producing them.

The theory was developed by the classic economists, including Adam Smith, in the late eighteenth and early nineteenth centuries. But it was Karl Marx who really latched on to it for his critique of capitalist exploitation.

There are a lot of things wrong with the labor theory of value—it ignores the importance of capital and the profit motive, for example—but simply put, it is a terrifically naive view of how the world works. See MARX, KARL.

Laffer Curve

Named after economist Arthur Laffer who, according to popular myth, drew the curve on a napkin while trying to explain the effect of confiscatory tax rates to a journalist. However colorful the story, Laffer did not invent the curve. It simply illustrates the commonsense notion of the trade-off between tax rates and tax revenue. The tax revenue curve itself sits like a hedgehog on the horizontal axis. If the tax rate is zero, then tax revenue—plotted along the vertical axis—is zero. As tax rates rise, so does revenue. When the rates become burdensome, then revenue starts to fall, and when the tax rate is 100 percent, revenue falls back to zero because people have no incentive to earn money. See REAGANOMICS.

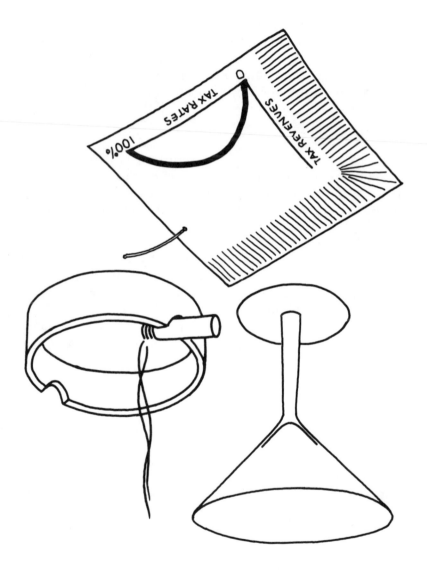

LAFFER CURVE

Lagging Indicators

A group of statistical indicators, like inventories, unit labor costs, or interest rates, used by forecasters to mark possible turns in the business cycle. Lagging indicators are the opposite of leading indicators because they trail economic activity. For example, when the lagging indicators start going up, it's thought to be a sign that economic activity has already peaked and a recession is on the way. As is true of all forecasting tools, some economists swear by the lagging indicators, others ignore them. See LEADING INDICATORS.

Laissez-Faire

A French phrase meaning, roughly, "leave it alone." First uttered by the Physiocrats and then taken up by Adam Smith and his followers, the phrase expressed frustration with government interference with the operation of the market. (If the government doesn't laissez-faire, then the invisible hand can't work.)

The phrase is usually applied to government regulation of economic activities; only the most extreme adherent would deny government's role in maintaining law and order, providing for the national defense, and undertaking certain public works projects, like utilities.

Land

One of the three factors of production, labor and capital being the other two. Land means more than earth. It means any natural resource, like minerals and water. As an input for production, land is different from labor or capital in the sense that it is fixed by nature—and depletable.

Leading Indicators

One of a group of measures of the country's economic health prepared by the Commerce Department. The leading indi-

cators consist of ten measures, including the number of hours worked in certain industries, claims for unemployment, new orders for consumer goods, stock prices, business and consumer borrowing. They are supposed to forecast the direction of the economy.

The leading indicators are published on a monthly basis. If they show smart monthly rises, then watchers feel reassured that strong economic activity lies ahead. If, however, the indicators fall three months in a row, they are supposed to foreshadow a recession. Supposed to. The leading indicators are correct only slightly more often than they are wrong; nonetheless, they are watched by economists, investors, and the press. See LAGGING INDICATORS.

Less-Developed Country

A euphemism for a Third World country, more faddishly called a developing country. That is, any poor country. LDCs are often found in Africa and Asia, rarely in North America or Europe.

Liability

A formal way of designating a debt. Liabilities are the opposite of assets: Assets are owned, liabilities are owed. On a balance sheet, liabilities equal assets—that's why it's called a balance sheet. See ASSET.

Liquidity

The ease or speed with which an asset can be turned into cash.

Liquidity depends on the nature of the market for any particular asset. Some assets, like shares in IBM or General Motors, are very liquid—owners can sell them for cash in a few minutes. Other assets, like homes or airplanes, are illiq-

uid—owners must wait awhile before buyers can be found. The only asset that is completely liquid is, of course, cash. A third-party check drawn on a bank in Algiers might be completely illiquid when the holder appears at a bank window in Des Moines.

A business is said to be liquid when it has enough cash to pay back its creditors. A business that is illiquid runs the risk of not being able to pay its expenses on a timely basis and, in fact, of going bankrupt. Hence, when a business goes bankrupt, and its assets are sold, the process is known as liquidation.

Long Run

Any period of time long enough to allow a firm to change or vary all its factors of production.

Say that Framistan International discovers that its orders outpace its production. And say that it takes Framistan International two years to build new production capacity. Since the only thing Framistan can do—in the short run—is to work the existing production facilities harder by using more labor and raw material, then the long term for Framistan is two years.

For economists, the term long run is very precise: The magic moment of equilibrium is what happens in the long run, and disturbances in equilibrium are, by definition, what happens in the short run. For example, unemployment is a disturbance of the short run and full employment is the equilibrium condition of the long run. (This is what Keynes meant by his famous phrase "In the long-run, we are all dead.")

But the term can also have a very unspecific meaning. If an investor says she is investing for the long run—that she is willing to wait a "long" time for her payoff—she can mean anything from two years to twenty. See SHORT RUN.

Loose Money

One of those colorful phrases that capture perfectly a complicated idea. Loose refers to the monetary policy undertaken by the Federal Reserve, and money refers to the amount of money in the economy.

When the Fed decides to make credit readily available to borrowers, it increases the supply of money. This is supposed to lower interest rates which, in turn, makes borrowing attractive. In other words, loose money means that the Fed has loosened its grip on the growth of the money supply.

The opposite of loose money is tight money. See FEDERAL RESERVE SYSTEM; TIGHT MONEY.

Mm

Macroeconomics

The part of economics that looks at the forest, rather than the trees (as opposed to microeconomics which looks at individual trees).

Macroanalysis deals with big-think categories, or what economists call the aggregates. For example, individual decisions to save or consume are lumped together, christened national savings or national consumption, and are fitted into a macroeconomic framework.

A classic framework—and one which appears with com-

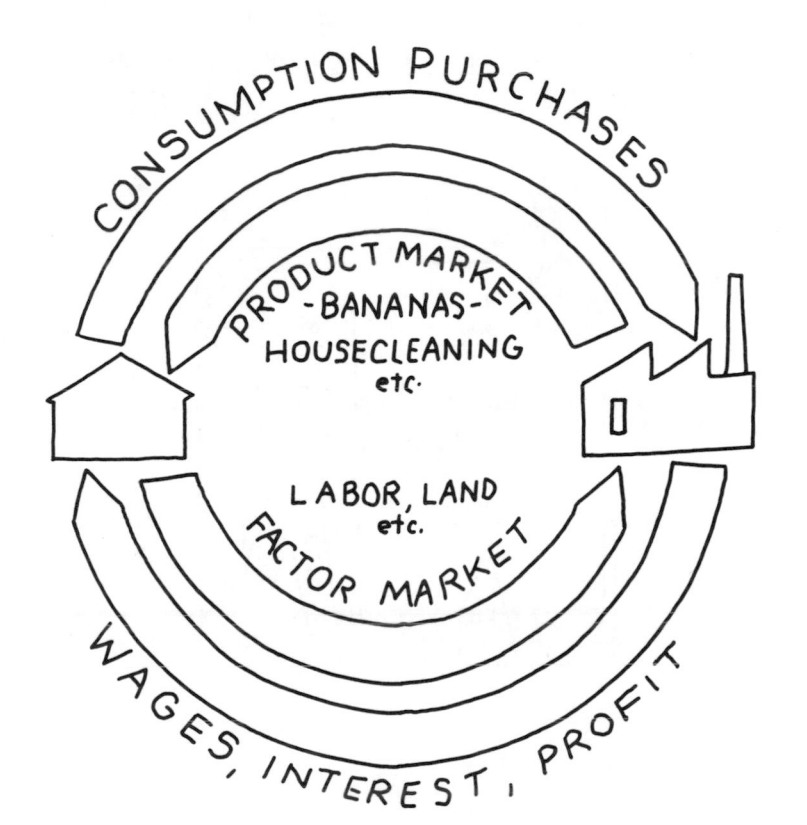

MACROECONOMICS

forting frequency in economics texts—is called the circular flow of income and products. The circular flow takes the form of a giant circle (what else?) with arrows that show the goods and services from households flowing to the factor market and then to firms, and then back again, flowing from firms to the product market and then to households; the income flows go the opposite way between households and firms. The circular flow is just a commonsense way of saying that households offer labor to firms and firms offer products to households, and households pay for products with money that comes from firms that pay households for labor.

Macro also looks at topics like fiscal and monetary policy, the rate of economic growth and unemployment, the balance of payments, and inflation. The point is to discover the relationships among them—for instance, a loose monetary policy causes inflation—and thus get a handle on how the economy, in general, works.

Just because macro deals with big ideas doesn't mean that it's more important than micro, however. An understanding of both these parts of economics is vital. See MICRO-ECONOMICS.

Malthus, Thomas Robert (1766–1834)
A British economist and first-rate gloom-and-doomer.

Malthus is known for his *Essay on the Principle of Population as It Affects the Future Improvement of Society* (1798), which essentially says there can be no future improvement of society because population growth will outstrip the food supply. (Although he did hold out hope in the form of pestilence, vice, famine, and war, which—by raising the death rate—could slow population growth.)

So far, Malthus's predictions have not been realized. He underestimated the capacity for improvement in agricultural productivity and the ability of people to limit family size. Nonetheless, there are still Malthusians who fear that his theory may hold for many Third World countries.

Marginal Analysis

The study of how small changes—increases or decreases—in the total amount of one variable affect one or more related variables. Granted, that sounds dismayingly opaque, but read on; knowing what happens at the margin is crucial to understanding economics.

A good way to grasp marginal analysis is to work through a problem, and the problem most people encounter first in economics' texts is marginal utility. Think of a marginal unit as an extra one, something on the top. If you have a dozen cookies, then the thirteenth cookie would be a marginal unit. And think of utility as units of satisfaction, or "utils," that are received from consuming cookies.

Say you eat a dozen cookies a week and say you receive 24 utils from these cookies. If you receive 22 utils from the first eleven cookies, then the marginal utility from cookie twelve is two. In other words, the marginal unit is obtained by dividing the change in total utility (two) by the change in units consumed (one).

But, now enter the thirteenth cookie. A new, marginal cookie. With thirteen cookies a week, your total utility goes up to 25, but your marginal utility from cookie thirteen drops to 1. Your satisfaction diminishes because, at thirteen cookies, eating another becomes less of a treat.

What other variables will this decline in marginal utility affect? Most likely, when you reach the point of declining marginal utility from eating cookies, you will buy more of something else, like ice cream.

Marginal analysis might seem overly fussy, but it gets its explanatory power from the fact that most economic decisions involve small changes: you do not decide to spend your entire income on cookies, but only a small portion of it. Thus, in the example above, marginal analysis indicates the point at which you will stop spending some of your income on cookies.

Marginal analysis is used for many things in economics. It can be applied to consumers and firms, prices and costs.

Thus, in addition to marginal utility, economists study marginal revenue, marginal cost, marginal propensity to save and to consume. Marginal analysis indicates when consumers reach maximum satisfaction and when firms reach maximum profits. (Another quick example: Firms maximize profits when their marginal costs equal their marginal revenues, that is, when the cost of producing one extra widget equals the revenue that widget brings in.) See UTILITY.

Market

Something that is created when buyers are in contact with sellers, when supply and demand interact to determine prices. (And it doesn't matter if these buyers and sellers are eyeing each other over counters, talking on telephones, or transacting through carrier pigeons; as long as they are doing business, that business is called a market.)

The market is a primo notion in economics. Through the free—or relatively free—interaction of supply and demand, markets play a central role in allocating resources. The mechanism by which this is accomplished is called market forces. While that might sound dramatic, it isn't—at least, not in any theatrical way.

Market forces are the sum of all the decisions, small and big, to buy and sell. Consumers decide to buy toothpaste or cars, business people decide to produce widgets or framistans, workers decide to work for banks or publishers, investors decide to buy stocks or bonds. While prices guide decisions, decisions also guide prices. (Low costs can lure producers into widget making, but if low prices lure a rush of people to buy those widgets, makers might then raise prices.) Thus, economists like to say that market forces set prices. And prices, in turn, cue society on how to use its inputs efficiently by putting values on output.

That is, anyway, how market forces are described in most textbooks. But now for the caveats. Prices are also determined by market structure, that is, by how much competition is pres-

ent in any particular market. For example, if there are a lot of buyers and only one seller, then the seller has enormous power over the price of the product.

More to the point, if the price does not reflect the value that society places on the product, then the market cannot allocate resources efficiently. These kinds of markets are known as market failures. They include monopolies and oligopolies, and they invite government regulation. The government acts, presumably, with the best of intentions. Nonetheless, when it steps in to set prices or otherwise remedy a market failure, it violates the function of the market set forth at the beginning of this entry.

Marshall, Alfred (1842–1924)

A British economist who started his professional life as a mathematician. Marshall was able to use his mathematical training to strengthen the foundation of the classical school of economics. In fact, Marshall really turned economics into a science.

Marshall was an awesomely important influence. He developed many of the ideas of the neoclassical school. And his textbook, *Principles of Economics,* published in 1890, dominated the teaching of economics for fifty years.

Marx, Karl (1818–1883)

German economist and political propagandist.

Marx and his longtime pal Friedrich Engels zapped the world in 1848 with their *Communist Manifesto.* It begins, ominously, with the words "A spectre is haunting Europe—the spectre of Communism. . . ." It goes on to warn the bourgeoisie that the ultimate product of capitalism is, "above all, its own grave diggers." And it concludes by exhorting the proletariat, who "have nothing to lose but their chains," to unite and overthrow the capitalist system.

Strong stuff. But it was downhill from there. In his *Critique*

of Political Economy and *Das Kapital,* Marx laid out what has come to be known as Marxian economics. But unlike the Manifesto, this prose is mind-numbing.

Essentially, Marx takes David Ricardo's labor theory of value as an absolute. That is, the amount of labor used in the production of goods determines their total value—forget any possible contribution from entrepreneurs (capitalists) and machinery. Marx argues that capitalists exploit labor by paying workers less than the value of their product and keeping this surplus for themselves as profit.

Unpleasant as this situation is, things are bound to become more unpleasant. According to Marx, the world is passing through stages on its inevitable way to communism. And class warfare is necessary for the capitalist stage to pass into socialism, and then to communism. Warfare? Inevitable? Capitalists will eventually try to maintain their profits by cutting costs through labor-saving innovations. This, in turn, will create an army of unemployed willing to undercut their working brothers and sisters by accepting lower wages.

In short, falling wages and an ever-worsening lot for the proletariat transform class struggle into class war. The ultimate victory of the proletariat comes when it seizes the means of production, abolishing private property; soon after, the state withers away.

Capitalism, however, has proved more durable than Marx anticipated. Most importantly, the past century has seen real wages, and living standards, rise steadily. The proletariat now has a lot more to lose than their chains—their microwave ovens, vacation cottages, and videocassette recorders, for example.

Further, capitalism itself has exhibited a wily flexibility. The twentieth-century version is really a mixed bag of private and state ownership. Capitalism has definitely passed through a stage. But while what has evolved is different from unfettered laissez-faire and not exactly unbridled government intervention, one thing is for sure—Communism it ain't. See LABOR THEORY OF VALUE.

Mercantilism

A reigning economic policy during the seventeenth and eigh-teenth centuries that organized trade to promote national wealth and power.

The guts of mercantilism was a system of export and im-port policies designed to ensure a balance of payments sur-plus. Simply put, countries tried to export more than they imported. Thus, exports were encouraged with government subsidies and imports were discouraged by high tariffs. Since international trade was settled in gold, successful mercantilist countries were able to keep a lot of gold in their treasuries.

MERCANTILISM

And the more gold a country had, the richer and more powerful it was.

Part of the obsession with gold was just that—an obsession. But part of it was also practical. Gold was a crucial strategic consideration in any decision to fight a war (a popular activity back then). Nations with plenty of gold could go to war because they had the means to pay soldiers, bribe neutral countries, or purchase war goods from abroad.

Perhaps the best example of how complicated and pervasive a successful mercantilist system could be was the arrangement Great Britain had with its colonies, including the American colonies. Through an involved set of laws, called the Navigation Acts, Britain controlled most of the exports and imports going to and from the colonies, and paid subsidies for certain goods grown in the colonies. In this way, Britain gave the colonies the role of producer of raw materials and took for itself the role of manufacturer.

The general mercantilist notion was that the colonies existed to make the mother country (remember, this was centuries ago when terms like mother country were considered quite okay) not only self-sufficient, but able to export its goods.

Obviously, mercantilism meant government-managed trade. And as such, it's not so different from the protectionism people jaw about today. More to the point, neither holds much appeal for economists who think that free trade is the best way to increase national wealth. See FREE TRADE; PROTECTIONISM.

Microeconomics
The part of economics that deals with the trees rather than the forest (as opposed to macroeconomics, which looks at the entire forest).

Micro is often called price theory because it studies how the decisions of individuals—a consumer or a firm—determine prices. That means that microanalysis also studies mar-

kets, resource allocation, and income determination. Rather big topics. Moreover, microeconomics reaches beyond those issues to include macro problems. That is, many macro issues rest on a micro base: Inflation, for example, depends on the pricing behavior of consumers and firms and how markets are organized, as well as on monetary policy.

Just because micro deals with the ground floor of economics doesn't mean it's less important then macro. Understanding economics requires understanding both these parts. See MACROECONOMICS.

Minimum Wage
Federal, state, or local law imposing a floor on wages.

This is one of those well-intentioned but utterly flat-footed gestures the government often makes when it interferes with the market. The notion got its start during the New Deal: The government set a minimum wage so the bosses couldn't exploit workers by making them labor for a pittance. It was, and is, targeted at the low end of the labor market—unskilled minorities who are particularly vulnerable.

The intention, of course, was to protect workers by ensuring a living wage. The result, however, is rather perverse. While the minimum wage does indeed protect workers, it only serves those who can find jobs at the artificially high wage. In other words, there may be people willing to work for less, but employers can't hire them legally. In fact, at higher wages, some employers must hire fewer workers. Many people believe that the minimum wage law contributed to unemployment among those who most need jobs—typically, unskilled black teenagers.

Mixed Economy
An economic system that falls somewhere between a totally free market and a government-directed one; that is, an eco-

nomic system that has characteristics of both capitalism and socialism.

Most economies of the world are mixed, but in wildly different proportions. Some Soviet bloc countries, for example, have a soupçon of capitalism, most often in their agricultural sectors. And some free market countries, like Hong Kong, have a dollop of socialism, mostly in the provision of services. The rest, like the United States, have a more equal portion of both.

Model

A theory that is supposed to capture the workings of the real world.

A model is no more than an idea or description about how certain things affect other things. If, for example, you think that eating chocolate eclairs every night will make you fat, you are using a model describing how chocolate eclairs affect your body.

Models can range from the simple (one equation with two variables) to the awesomely complex (hundreds of equations with thousands of variables—consumption, saving, income, wealth, and so on and so forth—that are used to express macroeconomic relationships).

Economists like to think of themselves as scientists because they use models to test hypotheses. For example, to test the hypothesis that consumption expenditures are a function of income, an economist would first build a model that reflects that relationship, using a mathematical expression of the relevant variables. In this case, consumption expenditures would be on the left side of the equation, income on the right side.

Next, he or she would test the model by attaching numbers (called the data) to those variables. In this case the numbers might be total income and total consumption spending over several years. If the results seem to be related in some non-

random way (and seem to correspond to what the economists think actually happens—more income means more consumption spending), then the model is thought to be sound.

When economists get hold of a sound model, they can use it to forecast the future. If, for instance, income is likely to go way up next year, then economists will forecast that consumption spending will increase, too.

Economic models are vulnerable to one general criticism: They cannot completely capture all the variables that interact in any real-world situation. Granted, sometimes this criticism takes the know-nothing form of "Models are too abstract, reality cannot be expressed by a bunch of numbers." The more sophisticated would agree that even though the soundest model in the world is a description that, necessarily, must be incomplete, models can nonetheless hone ideas and produce useful insights into how the world works. See ECONOMETRICS.

Modigliani, Franco (1918–)

Modigliani is both a superb economist's economist and one whose ideas have had impact beyond economists' circles.

In 1985, Modigliani won a Nobel prize for two of his contributions to economic theory. The first, known as the Life-Cycle Hypothesis, argues that people's savings behavior is aimed at providing enough money to maintain spending during retirement. In other words, people save to build nest eggs during their working years and then spend—or, stop saving—after retirement.

The second contribution, known as the Modigliani-Miller model, analyzes how a corporation's financial structure—its debt and equity—affects the value of its stock.

While he is decidedly a Keynesian (as opposed to a monetarist or a rational expectationist), Modigliani was one of the first Keynesians to accept the importance of monetary policy. See SAVINGS.

Monetarism

School of thought that holds that changes in the supply of money are the chief determinants of economic activity.

The most famous monetarist—and the one who has pushed its agenda with remarkable ferocity—is Milton Friedman. While the exact doctrine varies from monetarist to monetarist, Friedman's central role has made his brand the classic version.

The monetarist view of the way the world works can be simply put. Too much money in the economy leads to increases in the general price level (inflation) and to economic booms; too little leads to decreases in prices and to busts; and up-and-down changes in the money supply lead to an up-and-down economy. The effect of changes in the money supply is not immediate, however: There's a long, and uncertain, period of time before changes have an impact on the economy.

No surprise, then, that monetarists are utterly fixated on the Federal Reserve: For them, the Fed, as the author of monetary policy, is the most important player in determining the course of the economy.

But their fixation takes the form of severe vexation. Because, according to monetarist theory, the Fed can never be sure when its policies will bite, monetarists don't believe it has any business trying to manage the economy through its actions.

Thus, monetarism argues that the Fed should just aim at supplying the economy with money at a constant rate of growth. Indeed, if the Fed turned into a little robot, supplying new money at around 3 percent a year, monetarists would be well satisfied.

Since monetarists believe that fiscal policy is not very powerful (except for its capacity to make mischief) and that monetary policy should be run like a robot, it follows that they are against most forms of government interference in the economy. In fact, Friedman has translated his free market economic arguments into a political philosophy.

Monetarism is the chief ism ranged against Keynesianism. Where the extreme form of monetarism says that money supply is all important, the extreme form of Keynesianism says that fiscal policy is all important.

The two isms used to be sharply divided, if not downright antagonistic. During the Keynesian heyday in the late 1960s and 1970s, monetarists warned that the excess money creation would result in inflation. They were ignored. They were dismissed as naive. They were, however, right. And nowadays most economists think that *both* fiscal and monetary policy are crucial players. See KEYNES, JOHN MAYNARD; NEW CLASSICAL ECONOMICS.

Money
Any generally acceptable means of payment in exchange for goods and services.

Money replaced barter which, as a system of payments, was extremely awkward. With money, people can directly exchange their labor for money and then exchange their money for steaks or haircuts. That makes money incredibly more efficient than barter, where people might have to exchange their labor for bananas and then find other people who like bananas enough to be willing to trade their steaks for them.

Of course—and this is the point—even bananas can be money if everybody accepts bananas as a medium of exchange. . . . Then people with steaks would be happy to exchange their steaks for bananas, knowing that they could then easily exchange those bananas for haircuts. Individual taste for bananas would not be an issue.

At any rate, money can be anything that people are willing to accept as a medium of exchange—pieces of paper, wampum, or bananas. But money has two other roles: as a unit of account and as a store of value.

A unit of account means that the units in which money is measured—dollars and cents for paper money, and pounds

and ounces for banana money—become the price at which all goods and services can be valued. Thus, a steak can be said to be worth $10 or 20 pounds of bananas, a haircut worth $20 or 40 pounds of bananas. Furthermore, by having a unit into which everything can be translated, it becomes possible to combine various goods and services. For example, the prices of all the goods and services in the economy can be added together to produce one measure of economic activity, or gross national product. (Nothing more, really, than the old trick of combining apples and oranges.)

A store of value means that money can be used to warehouse wealth for future purchases. Sellers will accept money today, even if they have no immediate plans to spend it, because they assume it will keep its value over time. Bananas, of course, would make lousy money—they might rot away before they could be spent. Not that paper money is perfect. Inflation could erode the value of paper money just as surely as oxidation erodes the value of bananas. Just not as quickly. (Usually.) See BARTER.

Money Illusion
The psychological response to changes in the nominal value of money—usually wages—rather than to changes in its purchasing power.

The money illusion is used to illustrate the impact of inflation on behavior. Say, the price level goes up 3 percent a year, and say that wages, too, go up 3 percent a year. People suffering from the money illusion think that the increase in wages has increased their purchasing power by 3 percent, when, of course, the rise in inflation has just left it constant.

Thus, if the money illusion holds, workers might continue to work for the same wages during periods of inflation even though inflation is eroding their purchasing power and reducing their *real* wages.

Many economists think that the money illusion is itself an

illusion. They argue that people are well aware of the effects of inflation (and deflation) and respond accordingly, and quickly. See RATIONAL EXPECTATIONS.

Money Supply
All the money existing in an economy. But wickedly difficult to measure.

The narrowest measure of the money supply is called M1: It includes coins and currency in circulation and certain types of checking accounts. Next comes M2, which includes M1 plus certain types of small savings accounts. And then comes M3, which includes all the money in M2 plus the big money accounts held by individuals, corporations, and institutions.

The Federal Reserve tries to keep track of the money supply to control it. That is, the Fed estimates how much money is actually out there in the economy and how much money there should be, and then decides how much to supply. For an explanation of why the Fed should go through all this trouble—and it's more troublesome than this entry indicates —See FEDERAL RESERVE SYSTEM.

Monopoly
A market with only one seller.

If you are a seller, a monopoly is a nice thing to have. You can set a price higher than it would be under competitive conditions and rake in enormous profits. If you are a buyer, however, a monopoly is a nasty thing to face. You will have to pay that high price, and you may even encounter a shortage of product. In other words, high price and low output go together. The monopolist produces just enough to ensure plump profits, and that means less output and higher prices than would prevail in a competitive market.

Since—obviously—all sellers would like to have mono-

polies, why isn't the market dominated by monopolies? Two reasons. First, it is impossible to maintain a monopoly without what economists call barriers to entry. (If a business person sets up a monopoly in making framistans, the giant profits which accrue will only signal other business people to produce framistans, too. Unless, of course, the others are prevented from setting up shop.)

Barriers to entry can take several forms. If framistan production takes an immense capital outlay, for example, other —less wealthy—firms may face a financial barrier. Or, if making framistans requires specially skilled craftspeople, then other firms might face a lack-of-talent barrier.

Perhaps the most common—and most effective—barrier to entry is through legal restrictions. The firm that first discovered how to make framistans might enjoy a monopoly that comes from having an exclusive patent or copyright, for instance. Usually, however, the government grants monopolies on some notion of serving the public: Countries can give air carriers the right to fly certain routes, states can give power authorities the right to provide electricity, and cities can give a garbage collector the right to cart away garbage.

But what the government allows, it can also prevent. And that is the second reason why the economy isn't chock full of monopolies. When barriers to entry exist, the government— through its antitrust laws—can restrain companies from becoming monopolies.

Many firms, however, enjoy a certain amount of freedom in setting their prices without enjoying an actual monopoly. It is arguable, for example, that beer is all pretty much the same, and thus beer makers face a competitive market. It is arguable, but not exactly true. Beer makers have been able to differentiate their products so that certain brands command such fierce loyalty that makers can charge more without losing all their customers. This kind of market, which falls somewhere between monopoly and competition, is called monopolistic competition. See REGULATION.

Monopsony
A market with one buyer. (As opposed to a monopoly which is a market with one seller.)

Just as a monopolist has the power to set the price for his or her product, a monopsonist can set the price at which he or she will buy. Monopsonies are rarer than monopolies, but they do exist. In a company town where there is only one buyer of labor, for example, the employer has power over the wage level. (If a worker thinks his wage is too low, he has a choice—get on unemployment or get out of town.)

Multinational Corporation
A firm that operates in more than one country, usually through a wholly owned subsidiary.

Multiplier
A shorthand way of capturing a complicated process by which changes in spending affect the level of national income.

To get an idea of how the multiplier works, consider the following example of how a rise in investment spending works its way through the economy to produce a higher gross national product. Say that a firm increases its investment spending by $100 million by building a new factory. And say that this initial increase in spending shows up in wages paid to the factory builders. (It could also show up in higher profits, rents to the owners of the land, or interest payments.)

Some of this $100 million will be spent by the recipients, generating a second-round increase in national income. Say, for example, of the original $100 million, three-quarters of it, or $75 million, is respent. Thus, the initial $100 million has now yielded a $175-million increase in national income.

But it doesn't stop here either. This round of spending will result in more wages for other workers because the recipients of the $75 million will, in turn, respend three-quarters of it—

or $56 million—for a total increase of $231 million in national income.

And so it goes. Each round of spending will generate another round, as one worker's spending puts money in the pockets of another worker. By the time the increases are too small to measure, $400 million will have been added to national income. That is, national income will get a $400-million boost from the original $100 million. A quadruple bang for a buck.

In this case, the actual number attached to the multiplier is four. The size of the multiplier is based on a marginal propensity to consume, or spend, three-quarters of every dollar. Note that the flip side of spending is saving, so that a marginal propensity to consume three-quarters of every dollar means that one-fourth of every dollar is saved. (And the mathematically adept might also note that the multiplier is the reciprocal, or upside-down, of the marginal propensity to save.) This saving "leakage" stops the multiplier from generating income increases forever.

The multiplier is a mainstay in Keynesian fiscal policy. It's used to estimate the impact that increases in government spending or tax cuts will have on national income. Despite this seeming precision, however, multipliers have proved, in practice, to be less than precise. See MARGINAL ANALYSIS.

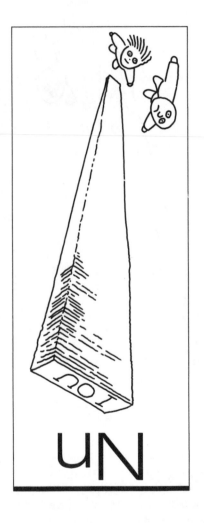

uN

National Debt

The federal government's total outstanding debt. (Note that this is different from the federal budget deficit, which is just the debt incurred by the government over one fiscal year.)

In a world of balanced budgets where government's revenues equaled its expenditures, there would be no national debt; that is, the government would have no need to borrow.

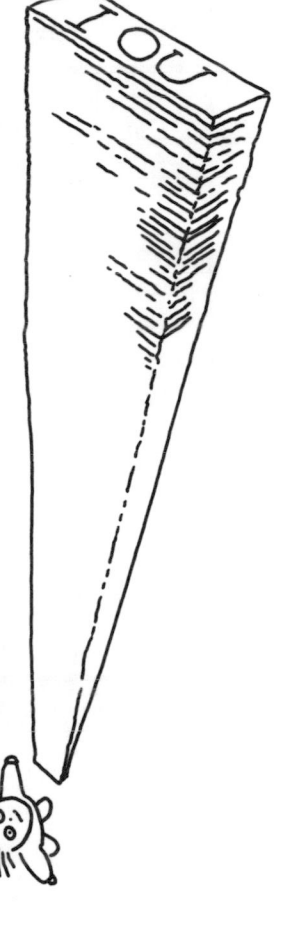

NATIONAL DEBT

In the real world, however, government has been spending more than it takes in, and thus it has been borrowing to cover the difference. The total of these budget deficits amassed over the years is called the national debt, and amassed is the right word for it. The absolute size of the debt is over $2 trillion.

A growing debt means a growing need to get somebody to pay cash money to service the interest on that debt, if not to pay back some of the principal. There are two ways to get this money: The government can print some up, courting a bout of inflation; or it can increase taxes, courting a slowdown in economic activity. Neither route is appealing.

Nonetheless, there is another—less alarming—perspective from which to view the debt. First, the debt is something we owe ourselves. Interest payments on the debt, as long as they are paid to U.S. citizens, are like transfer payments. (Follow this: The government borrows money by exchanging its debt securities—Treasury bonds, for example—for cash money, some of which goes to pay interest on previously issued debt securities.)

Second, sheer size is a little misleading. The national debt has grown, but so has the economy. Indeed, when expressed as a percent of gross national product, the debt actually declined during the 1960s and early 1970s (though it started to rise again during the 1980s). And finally, inflation—which always favors debtors over creditors—has reduced the value of the debt. For example, when the $1.6 trillion of 1984 debt is adjusted for inflation, a full $4 billion disappears.

National Income

National income is another way of saying gross national product. It designates the money flow from the total earnings of all the factors of production. National income includes wages and salaries, interest, rental income, and profits.

Keeping track of national income is the province of the Department of Commerce. In fact, the statistical collection of

all sorts of measurements taken of the economy is called the National Income and Products Accounts. Most of the series in NIPA are published monthly.

Nationalization; Nationalized Industry

Nationalization is what happens when government takes over a business or industry that had been owned privately, usually by stockholders. A nationalized industry is what results from nationalization.

Whether nationalization is an act of stupidity or enlightenment is a matter of hot debate. Supporters of nationalization argue that it creates the opportunity to realize economies of scale and/or that it removes the opportunity for a monopoly to gouge the public. (These are economic reasons; occasionally, however, government seizure is an act of pure politics, done to eliminate foreign ownership.)

Critics, on the other hand, argue that nationalization leads to inefficiency because state industries generate large, costly, and wasteful bureaucracies.

State ownership of large sectors of the economy—such as telecommunications, banking, power generation, and railways—are usually the province of socialist and less developed countries. But most countries indulge in some form of nationalization.

Negative Income Tax

An idea that is part of the public policy debate on how to make sure that poor people don't go hungry and homeless without, at the same time, making them permanent welfare recipients.

Roughly, a negative income tax would work like this: The government would determine a poverty line—a point under which family income would not be allowed to slip—and then pay whatever amount brings people to that line. (If a family earned no money, for example, they would get the full trans-

fer payment.) Taxes would be levied on a progressive basis only on income above the poverty line.

The negative income tax, which is actually an income subsidy, has several virtues. First, with a certain minimum income ensured, people would still have incentives to earn money because they would keep some portion of their earnings above the poverty line. And second, it could replace the current unwieldy, inefficient, and expensive welfare system.

While the idea has support from both liberals and conservatives alike, it was the brainchild of conservative economist Milton Friedman.

Neoclassical Economics
A school of thought that, obviously, branched off from classical economics.

Early names here include William Jevons (British), Karl Menger (Austrian), and Léon Walras (French); but perhaps the most famous neoclassical economist was Alfred Marshall (British). While neoclassical thought reflects the classical emphasis on the importance of markets and competition, it expresses those relationships with sophisticated mathematics and statistics.

Net National Product
Gross national product minus depreciation. See DEPRECIATION.

Net Worth
The value of a company after total liabilities are subtracted from total assets. If that number is positive, then the company is said to have a positive net worth. If the number is negative, then the company is said to have a negative net worth—a condition that usually leads to bankruptcy.

New Classical Economics

School of thought that holds that the market produces the best solutions to economic problems.

Granted, that definition sounds just like the definition for classical economics, and granted, the new classical economics (NCE) does take strength from some of the same ideas; nevertheless, there are important differences. New classical economics really is new. It's equipped with the latest mathematical techniques, a bunch of theoretical refinements, and an undeniable sophistication. In fact, NCE has emerged as a major alternative to Keynesian and monetarist thinking.

Most importantly, NCE takes the pivotal assumption in microeconomics—that markets work—and applies it to macroeconomics. On the face of it, this might not seem to be a major achievement. Nonetheless, the chief theoretical embarrassment for Keynesian economists is that the assumptions they believe to be true in micro cannot be translated into their macro framework.

For example, a key tenet in microeconomics is that the market equilibrates supply and demand through prices. (Fancy talk for saying that prices will adjust so that buyers and sellers can make transactions.) Keynesian macro theory, however, says that prices, especially wages, are sticky—they don't adjust. And since some prices don't fall in response to falling demand, government must step in to replace private demand by public spending.

NCE also translates the micro notion of maximization, or rational expectations, into the macro realm. Microeconomics assumes that people maximize by behaving to further their own best interest. By assuming that the same holds in the macro realm, NCE argues that people react to changes in the economy by protecting themselves; that is, people will anticipate government shifts in policies in response to economic changes. Thus, it's difficult for the government to surprise its citizens into behaving against their own best interests. Again, this is different from the old Keynesian macro notion

that the government, by manipulating people's behavior through its policies, can manage the economy.

Needless to say, NCE goes on to argue that active government management of the economy is, at best, ineffective; at worst, government management makes things worse. Further, NCE places a great emphasis on the power of monetary policy—particularly its power to cause mischief like inflation.

In terms of policy, then, the bottom line for the new classical economics is that the government should not interfere with markets, that both fiscal and monetary policy should be rather passive and predictable. See KEYNES, JOHN MAYNARD; MONETARISM; RATIONAL EXPECTATIONS.

NIPA. See NATIONAL INCOME.

Nonprice Competition

A way of competing for buyers without lowering prices.

Say that Framistan Perfume wishes to increase its market share without lowering its price. Framistan can do several things. It can "differentiate" its perfume by changing its packaging or quality; it can advertise that its perfume provides the wearer with superior sexual appeal; or it can offer buyers a free tote bag with each purchase. While these strategies might sound a bit hokey, they are usually effective.

Nonprice competition often results in more varieties, or styles, of product being offered to consumers.

Normative Economics

The usual definition of normative economics says something about its being the branch that inquires into "what ought to be. . . ." What goods and services ought to be produced, who ought to own resources, how income ought to be distributed. This is just an upstanding way of saying that economic ques-

tions, or policy issues, involve value judgments. In other words, what's called normative economics is what happens when people pronounce on what is good and what is bad.

But value judgments depend on one's perspective. An unemployed, unskilled worker has a different view of how tax policy affects income distribution than does a boss or an heiress. Thus, a more accurate definition of normative economics would be: The phrase applied to the moral side of public policy debates. See POSITIVE ECONOMICS.

Oligopoly

A market that is dominated by a few large firms.

In terms of market structure, an oligopoly falls between monopoly (one producer) and competition (many producers). In terms of the firms' practices, however, an oligopoly can transform itself into either one. If the firms agree to set prices, an oligopoly becomes a monopoly; if the firms enter into a price war, an oligopoly becomes competitive.

The incentive for becoming a monopoly by price-fixing is clear: All the firms can raise their prices and enjoy larger profits. In the competitive case, however, the incentive only operates for the first firm, which, when it cuts its price, can grab a bigger share of the market. (After that, of course, the rest of the firms follow suit and the ensuing price war will reduce everybody's profits.)

Despite these incentives, oligopolies—and their prices—usually remain stable. For two simple reasons: If firms collude on prices, they will be in violation of the antitrust laws; and if they engage in a price war, their profits will suffer. Since either outcome is not terribly desirable, the result is

OLIGOPOLY

that oligopolistic firms will keep their prices close together and compete, instead, in other ways. Particularly by advertising. (Just check out the money that oligopolies—from carmakers to beer brewers—spend on advertising aimed at touting product characteristics other than price tags.)

Oligopolies are not desirable. They can set prices higher than might be allowed under competitive conditions (and generally show higher profits despite the tons of money spent on advertising).

Open Market Operations

Name given to the activity of buying and selling government securities when it's done by the Federal Reserve System.

Open market operations are set into motion at the monthly meeting of the Federal Open Market Committee (FOMC). At these gatherings, the governors of the Federal Reserve in Washington, along with presidents of five regional Federal Reserve Banks, vote on how to conduct monetary policy over the near term. Just what goes on at these meetings is a big secret until, weeks later, the minutes are published.

The operations themselves consist of the Fed buying or selling government securities. If, for example, the FOMC decides that the economy is growing too fast—and inflation threatens—it takes securities from its own portfolio holdings and sells them on the open market.

The buyers pay for the securities by giving the Fed checks drawn on their banks. When the Fed presents those checks for payment, an interesting thing happens. The banks lose some of their "balances with the Fed," which means that their reserves are reduced. When bank reserves (assets) are reduced, the banks must, of course, reduce the amount of their deposits (liabilities). And, since the banking system is a fractional reserve system—where reserves back only a fraction of loanable money—fewer reserves means a whole lot fewer deposits or bank-created money.

In other words, the sale of government securities reduces

the amount of loan money available from banks, which means there is less credit in the system. Less credit, in turn, means higher interest rates. Higher interest rates, in turn, means less business activity. Less business activity, in turn, means slower economic growth.

There is, thus, a rather long chain of events between the FOMC decision that the economy is overheating and the actual slowing of the economy. And in a big, complicated economy, a long chain of events leaves a lot of room for random mess-ups. Simply put, policy is one thing, results another. Nonetheless, open market operations are considered to be the most powerful tool the Fed has in its management of monetary policy. See BANKS AND BANKING; FEDERAL RESERVE SYSTEM.

Opportunity Cost

A measure of what could have been. The amount of money given up when resources are used to produce one thing instead of something else.

Opportunity cost is a classic kind of economist's notion. For an accountant, for example, a cost is a cost; it is whatever must be, or has been, paid. For an economist, however, a cost is mixed up with another notion, one about scarce resources. That is, the real cost of using scarce resources to produce a tractor is the value of other things that cannot be produced when those resources are used to produce the tractor.

By this logic, everything has a cost. Even leisure time—the cost of spending an afternoon playing tennis is what you could earn working.

Opportunity cost is useful in deciding between two alternatives. The usual example given in economics texts is about a self-employed person who could earn, say, $40,000 a year as an advertising executive and chooses instead to open a doughnut shop. If the doughnut maker takes home $20,000, he or she is $20,000 in the hole, when the opportunity cost of his or her alternative labor as an ad exec is considered.

Pareto Optimality

Named after Italian economist Vilfredo Pareto (1848–1923), pareto optimality represents a kind of a magic moment when resources are being used so efficiently that nobody can be made better off without making somebody else worse off.

As a specific notion, pareto optimality has a certain power because resources are usually being used inefficiently. Thus, any remedy will improve the welfare of some without damaging the welfare of others; optimality is something to aim for.

Planned Economy

Many definitions are possible (and accurate) here, but roughly a planned economy is one that eschews the price mechanism as the main way to allocate resources.

Typically, it is government that does the planning; that

PLANNED ECONOMY

is, the government decides on what and how things are produced, where investment money flows, and how income is distributed. Typically, too, government owns large chunks of the economy.

While the goal of a planned economy is worthy—to maximize the social welfare of all members of the society—the economic costs are usually big. Most obviously, government allocation of resources is dramatically imperfect. Granted, such allocation is meant to be inefficient in the sense that government decides on questions of social welfare first and economic ones second; nonetheless, the creation of a bureaucracy to oversee all economic matters often results both in bad decisions and in bad decisions that are permitted to persist.

All nations indulge in some forms of planning, but the most extreme forms are found in communist nations where the stated purpose is to tolerate only the barest hints of capitalism.

Positive Economics

A term indicating economic statements that are purely descriptive (versus normative economics, which is loaded with value judgments).

Positive economics is concerned with explaining how and why we got where we are, and what the outcome of a particular policy is likely to be. It does not concern itself with questions of whether where we are is good or bad, or whether the outcome of a policy is socially desirable or not. For example, in debates over tax policy and income distribution, positive economics will focus on questions of efficiency rather than equity. See NORMATIVE ECONOMICS.

Price System

Way of allocating resources that also transmits information on demand-and-supply conditions. That's the short definition. A longer, more satisfying, explanation requires a play-

by-play account of what happens in an idealized free market where buyers and sellers are free to interact.

Consider the simple island economy so beloved by economists. The scene opens on two sellers, one with bananas and one with oranges. They meet and greet as sellers but quickly turn into buyers when they exchange goods—let's say four bananas are exchanged for two oranges. (Note that prices don't have to be fixed in monetary terms, just in relative value: Two oranges are worth four bananas.) Scene two is a terrible storm which destroys most of the banana crop but leaves the orange crop—which has been harvested and stored —untouched. Scene three takes place a week later, after the banana famine has been felt by the island. This time, when the banana seller and the orange seller meet, they exchange two oranges for one banana. Final curtain.

What has happened? The relative scarcity of bananas has caused their price to rise; that is, bananas will now go to those who prize them enough to pay more oranges for them. And, voilà, the effects of the terrible storm have been transmitted through the price system.

Needless to say, the working of the price system in the real world is not as efficient as it would be in a simple island economy. In the real world, government regulations that set price floors (supports) or price ceilings (controls), and the existence of a certain class of goods known as "public goods," block the price system from the efficient allocation of resources and the transmission of information.

Say, for example, the government of the simple island economy institutes a price support program for oranges; that is, the government decrees that the price of oranges may not fall below two bananas. Thus, the orange seller cannot offload his oranges at the one banana price asked by the banana seller. Instead, the government must stockpile the unsold oranges—hardly an efficient allocation of oranges and certainly a misrepresentation of the supply-and-demand situation. (That, by the way, is how the U.S. farm program, with its complicated system of price supports, works.)

Private Sector

Economic activity—personal and business—excluding the government's contribution.

Privatization

A newly fashionable word to describe the process whereby the government transfers responsibility for delivering certain services to the private sector. The opposite of nationalization. See NATIONALIZATION; PUBLIC GOODS.

Productivity

A widely used term that can mean almost anything.

Even if the term is used precisely—a ratio that measures

PRODUCTIVITY

how much input produces how much output—its application can be fuzzy. Take the most often used measure of productivity, labor, which is usually expressed as the ratio of output to the number of labor hours spent producing that output. Labor productivity is said to be rising if it takes less time for workers to produce the same amount of output. But what does that really measure? Any number of things: workers are working harder, workers are working smarter, or workers are working with better machines.

At any rate, the measurement of productivity is less important than its result. The more productive an economy is, the better off its members are either because the same amount of goods and services are being produced with less work or more goods and services are being produced with the same amount of work.

Profit

The term seems straightforward enough, and in everyday life it is. But, for economists, the idea of profit is divided into two types—normal and pure.

Normal profits are embodied in the minimum return necessary to keep the factors of production doing what they're doing. It's the earnings that the owner of a business retains after she pays her explicit costs: rent to the owner of the store or factory, the salaries or wages to her workers, and the money for any goods or services she uses to produce her product. The money that's left after explicit costs are paid is then attributed to implicit costs: the return on factors such as the capital she has tied up in her business, and the opportunity cost of her labor.

Pure profits are what, if anything, is left after deducting all the explicit costs and implicit costs. These are returns above what is required to keep the factors doing what they are doing. If normal profits are like cake, then pure profits are like icing on the cake.

There are several ways of looking at profit—other than as theft by greedy, exploitative capitalists. First, profit can be seen as the return for bearing risk. After all, business is subject to failure; profits are the reward for taking a chance. Second, as Schumpeter argued, profit can be seen as the return for innovation. It's both the incentive that motivates entrepreneurs, or innovators, and their reward.

Third, profit can be, simply, a matter of luck. Consider the "windfall gains" captured by owners of oil or housing during the 1970s. And finally, profit can be a return to the owner of a monopoly. Here, monopolies include things like a patent on an invention, an outstanding tenor voice, some acres of land in Bordeaux as well as ownership of the company store in a coal-mining town. See WINDFALL PROFITS.

Protectionism

All-purpose term to describe government efforts to shield the domestic economy from competition from abroad. Protectionism can take many forms: quotas or increased taxes on imports, and/or subsidies and reduced taxes for the home team.

Protectionists like to identify situations that, they argue, merit special consideration. Some industries, like those that feed into national defense and agriculture, are said to have strategic importance for the country's ability to wage war. Other industries, like so-called infant industries, are thought to deserve protection until they have developed a requisite maturity—until they're big enough to fight their own battles. Of course, once they're big enough, or so goes a fourth argument, they become so important for the national economy that they, too, must be protected.

There are, however, many economic problems with protectionism. It violates the notion of comparative advantage which holds that all countries are better off if they produce whatever they are relatively best at producing. It can be ex-

PROTECTIONISM

pensive for consumers who are denied cheaper imports; likewise it can feed inflation because sheltered industries can inflate their prices. And, once protective legislation is in place, it's tough to dismantle. See QUOTA; SUBSIDY; TARIFFS.

Public Goods

Two parts to the definition here: Public goods can be consumed by more than one person and, once having been supplied to one person, cannot be denied to another.

This is an idea that demands concrete illustration: National defense is a public good—all citizens can use it and it is impossible to provide it for you without providing it for me, too. Other examples are public parks, clean air, and streetlights.

The example of choice for economists is lighthouses. While it's certainly true that a ship in a storm-tossed sea benefits from a lighthouse, the shipowner has little incentive to build and operate a lighthouse because, once it is beaming, all ships on the storm-tossed sea can use its signal without paying for it. In other words, while everybody benefits from the service, nobody wishes to underwrite its provision. Thus, it makes sense for the government to step in, build and operate a lighthouse, and either pay for it with general revenues or by taxing all shipowners.

However, solving the problem of who should provide public goods (note how appropriate their first name) only gives way to the next problem—how much of any public good should be provided. The answer is, unfortunately, that absent the free market, with its price signals, the government ends up guessing. Hence there is often inefficiency in the provision of public goods.

Too, some economists argue that the government has mistaken some private goods for public ones. And there are, in fact, lots of examples of public goods that are also provided by the private sector: garbage collection, mail delivery, education, housing, and police protection. Critics want these ser-

PUBLIC GOODS

vices "privatized"; that is, returned to the market so that price signals can ensure more efficiency in the delivery of these services. See PRIVATIZATION.

Public Sector
That part of economic activity that includes federal, state, and local governments, but excludes personal and business activity.

Qq

Quantity Theory of Money

A tautology that looks like this: $MV \equiv PQ$, where M is the nation's money supply, V is velocity (how fast money changes hands), P is the price level, and Q is the nation's output.

This tautology is the foundation for monetarism. In its crude form, the V and the Q are just passive, and P responds to changes in M. That is, if the money supply doubles, then prices double.

Sounds simple, and it is. It is also not quite accurate. Historically, changes in the money supply have not caused a one-for-one change in the price level because the two other variables, velocity and output, are not really passive. For example, when velocity falls, prices do not rise as fast as the money supply.

Anyway, a little looseness in the equation doesn't take away from the essential fact that sustained increases in the money supply will, eventually, push up the price level. See MONE-TARISM.

Quasi-Rent

Any extra revenue, after costs have been taken care of, that temporarily accrues to a factor of production. Generally, quasi-rents accrue to factors of production that are scarce for one reason or another.

Say, for example, somebody discovers that unbelievably great bananas can be grown on the island of Manhattan. During the time it takes imitators to pull down skyscrapers and cultivate bananas, the initial grower will reap a quasi-rent from his possession of great banana-growing soil. See RENT.

Quota

A quantitative restriction.

Quotas have three applications. When they are applied to imports, they become part of the nation's protectionist policy (as in the case of the United States slapping quotas on Japa-

nese autos). When they are applied to output they become part of a cartel's machinations (as in the case of OPEC which gives each oil-producing member a limit on the oil it can pull out of the ground). And when they are applied to an industry by the government, they become part of public policy (as in the case of the U.S. farm program).

It really doesn't matter for which of the three reasons quotas are used, the result is always the same—prices go up. In the first case, both the protected industry and the quota-levied foreign industry can raise prices on their products. And in the second and third cases, restricting the amount of available product (supply) pushes prices up.

QUOTA

Rr

Radical Economics
School of thought that is critical of mainstream economics—Keynesianism, monetarism, and new classical economics.

It's not a stretch to say that radical economics is anticapitalism. It is concerned with issues like the distribution of income and wealth, rather than with questions relating to economic growth.

Other than ideology, the chief difference between radical and mainstream economics is that the radicals include social and political issues as proper subjects of economic inquiry. They argue that traditional economics has become too focused and mathematical, and thus trivial. For their part, mainstream economists accuse radical economists of possessing too much cant and too little science.

Rational Expectations
A school of thought that describes how people respond to economic events. Dubbed ratex, it is based on the notion that people aren't fooled by government schemes to manipulate their behavior. Briefly, here's the reasoning: Since people have the same access to information that the government does, they can anticipate the government response to that information. Moreover, by anticipating government action, people can render government's policies ineffective.

Consider what happens when the budget deficit becomes alarmingly huge. People know, based on its past actions, that government will try to reduce the deficit by increasing taxes to raise revenue. However, since people act in their own best interest, they will protect themselves from a tax hike (and a fall in their after-tax income) by saving more money. But when they save more, they spend less which, in turn, slows the economy and depresses tax revenues. The net result, then, is that tax policy will be checkmated. The government ends up with less revenue, not more. And, of course, the budget deficit will grow larger.

Hence, ratexians argue that fancy government schemes to

"manage" the economy—that is, to manage people's behavior—are ineffective. Indeed, most ratexians would argue that government intervention in the markets causes confusion and second-guessing and can be downright dangerous. In short, they say that the economy would be better off if the government kept its hands to itself.

Doubtless, ratex represents a twentieth-century version of Adam Smith's argument that free markets create national and economic health more effectively than does government meddling. And doubtless, too, the ratexian notion that people can't be fooled by government policy seems little more than common sense. Nonetheless, ratexian notions seem fresh compared to the orthodoxy of the past twenty years when economists and politicians believed that government intervention in the markets could increase production, decrease unemployment, and control inflation. In the sixties, for instance, Keynesians thought that government could even smooth out the ups and downs known as business cycle fluctuations, and thus put an end to recessions; this was known as fine tuning.

Ratex got its start, slowly, in the sixties. It gained adherents among young economists in the seventies when persistent economic slumps put an end to grandiose notions like fine tuning. In fact, ratexian ideas even began to slip out of university economics departments to affect government policy (the effort to deregulate the economy that began in the Carter administration reflects this new laissez-faire thinking). And some of its adherents are looking to the eighties and nineties for an ideological coup that would end active government involvement in the economy. See KEYNES, JOHN MAYNARD; MONETARISM; NEW CLASSICAL ECONOMICS.

Reaganomics

An economic policy followed by the Reagan administration from 1980 on. The idea was that cutting tax rates on individuals and businesses would encourage them to save and invest;

this increased saving and investment, in turn, would generate higher incomes, more business, and more jobs. Reaganomics further argued that the super economic growth unleashed by the tax cuts would result in more tax revenues—enough to reduce the deficit and even produce a balanced budget.

That was the theory, anyway. The reality depends on who's doing the talking. Nobody disputes the fact that—despite astounding economic growth in 1983–84—the Reagan administration was dogged by truly huge budget deficits. But supporters fix the blame for those shortfalls on the Federal Reserve Board. The Fed, which had become concerned about runaway inflation, started restricting the amount of money in the economy in 1979, and that precipitated a steep recession. That recession, say supporters, ballooned the budget deficits by shrinking personal and corporate tax revenues and swelling government spending on programs like unemployment. Critics of Reaganomics, however, argue that the deficits were the result of too much tax cutting and too much government spending on nonrecession-related things, like defense.

The policy was dubbed Reaganomics because it appeared to be rather unique. True, Reaganomics directly contradicted policies followed by the three preceding administrations, policies that involved higher taxes and resulted in low rates of saving and investment and slow economic growth. Nonetheless, Reaganomics was not really revolutionary. The idea of stimulating the economy through tax cuts was enunciated by Keynes and advocated by his followers during the sixties.

Real
Value adjusted for inflation. What isn't called real is called nominal.

Real appears before words like gross national product, wages, income, and interest rate. "Real wages," for example, measures purchasing power. If the rate of inflation was 5 percent last year, and your wages went up 5 percent, then your purchasing power has stayed the same.

Real Economy

Here, real refers to the goods and services part of the economy as distinct from factor payments, or the money part.

Recession

Informally, any period of time during which economic activity is blah; formally, when the economy has not grown for at least two consecutive quarters.

Recessions can be mild (when they don't last too long and economic activity doesn't fall off too much) or they can be severe (when they last over a year and gross national product takes a dive). When a recession lasts a very long time and economic activity plunges, then it is called a depression. See DEPRESSION.

RECESSION

Regulation
Catchall term for government control or overseeing of various activities in the private sector.

The chief economic justification for regulation—and the usual textbook example—has to do with natural monopolies (situations where it's difficult to introduce competition).

Consider the electric power industry where there are tremendous economies of scale associated with having one single producer and distributor of power. A quick and dirty example: Think about the waste of resources in laying and maintaining two, three, or four sets of power lines.

In this case, the first thing the government does is grant a monopoly to one power authority for a geographic area. But then the government must do a second thing—it must regulate that authority to make sure it doesn't exploit its monopoly position by charging high prices. That task is usually assumed by a regulatory commission. Indeed, the toughest, most important, and most interesting issue for regulatory commissions is how to set prices.

On the one hand, the commission must set prices low enough so that the public isn't gouged; on the other hand, the commission has to set prices high enough to ensure that the utility gets enough revenue to stay in business.

But even if regulatory commissions could hit the price right, the right price in January might be totally wrong by November. An unexpected change in the price of an input, like fuel, or a change in the rate of inflation, can skew the whole cost structure and throw prices out of line. This problem is known as regulatory lag.

And these lags can be wicked. Commissions charged with representing the public interest are like giant ships—they cannot adjust quickly. They must call hearings and listen to endless testimony on all sides of the issue; they must deliberate and consider. Of course, while they're busy fulfilling their responsibilities, economic damage can mount.

Regulation, of course, encompasses more than just natural monopolies. Much, much more. In one fashion or another,

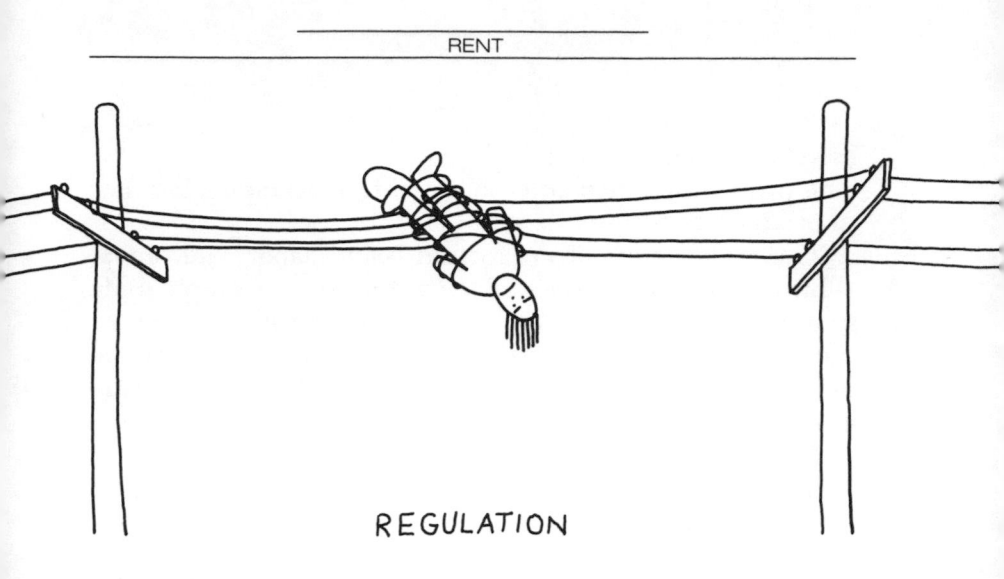

REGULATION

government fools around with almost everything. And, in one fashion or another, the rigidities and inefficiencies associated with regulation impede economic growth.

There are, at any rate, a lot of problems associated with regulation. These problems have led economists to take a new—and severely critical—look at the whole enterprise. And the fresh look has yielded some rather big changes. In some cases, businesses that were thought to be natural monopolies, like telecommunications, have been deregulated. In other cases, regulations that have created cartels, like the airline and trucking industries, have been dismantled. In fact, moves to deregulate the economy have become so commonplace, that deregulation has its own entry. See DEREGULATION.

Rent

The earnings accruing to a factor of production (land, labor, or capital) for which the supply is fixed.

Obviously, economists use the word rent differently than do landlords or tenants. Although land is the classic example of a factor earning rent, a superbly talented chef or an opera diva earns rent, too. See QUASI-RENT.

Resource Allocation

The process by which an economy directs its inputs in order to produce its outputs.

The basic factors of production—land, labor, and capital —have two characteristics. They are in limited supply (only so much land, so much labor, and so much capital), and they can all pretty much be put to more than one use (land can be used for an orange grove or a shopping mall). Thus, the way in which these scarce resources are used is a matter of some concern and not totally straightforward.

In some economies, the government simply decides who

RESOURCE ALLOCATION

gets what resource and for what use. State-planned economies, for example, might direct land, labor, and capital into
the production of industrial goods. In other economies, market forces decide who gets what. In a capitalistic economy, for
instance, the market might signal that land, labor, and capital
will earn their highest return producing consumer goods.

At any rate, economics is generally taken to be the science
that studies the intricacies of resource allocation.

Returns to Scale

Term used to describe what happens to a firm's output when
its inputs are increased. Returns to scale come in three varieties: increasing, constant, and decreasing.

Increasing returns to scale is what happens when a firm
doubles, say, the amount of land, labor, and capital it uses to
produce framistans, and, as a result, the amount of framistans it can produce trebles. (The firm has come out ahead.)
Constant returns to scale is what happens if a firm doubles its
inputs, and its output of framistans then also doubles. (The
firm is even-up.) And diminishing returns to scale is what
happens when inputs are doubled but output only increases
by a half. (The firm should go back to the drawing board.)

RETURNS TO SCALE

Ricardo, David (1772–1823)

An exceedingly influential classical economist (and a fabulously successful stockbroker).

Ricardo is known for *The Principles of Political Economy and Taxation* (1817) which set forth the rather gloomy conditions under which economic growth would eventually stop and wages would fall below subsistence levels. Nobody pays much attention to those arguments anymore; nonetheless, Ricardo's explanation of comparative advantage is still an important idea in international trade. See COMPARATIVE ADVANTAGE.

Ss

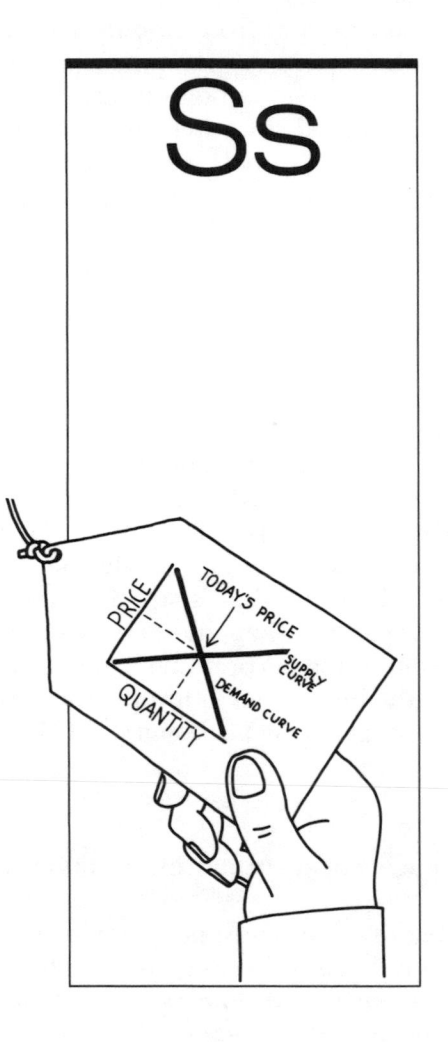

Samuelson, Paul A. (1915–)

Nobel prize–winner in Economics in 1970, for whom no superlative can be said to be an exaggeration.

Samuelson's contributions to economic theory and analysis —particularly to Keynesian economics—are both broad and deep. And surely important enough to give him the kind of stature that can be called towering. But most people know Samuelson through his textbook, *Economics,* first published in 1948. It has brought him lots of money, fame, and—above all—influence. *Economics* taught Keynesian economics to generation after generation of college students. Samuelson has taught at MIT since 1940.

Satisficing

A decision-making process in which the goal is to find a satisfactory solution, not the best, or optimum, one.

Satisficing runs contrary to the microeconomic assumption that firms seek to maximize profits. That is, when faced with a problem, businesspeople might decide that a satisfactory solution is better than one that is strictly profit-maximizing because, for example, the time and energy necessary to discover the best solution is beyond their immediate resources. The word—and the idea—were authored by Herbert Simon, who won the Nobel prize for Economics in 1978.

Savings

Whatever income is not spent on consumption is considered savings.

Thanks to Ben Franklin, most people think of saving as an act of thrift and little else—pennies going into a piggy bank or dollars under the mattress. The economists' view, however, is much more dynamic. Here, saving represents funds available for investment; saving provides funds to replace worn-out capital and to create new forms of capital. That is, saving underwrites the capital stock of the country without which

economic growth would come to a halt. Put another, more homely way, if a farmer (society) consumes his seeds (capital) instead of planting them (saving), eventually he will run out of food.

There are three sources of savings in the economy: personal, business, and government. Personal (sometimes called household), is the most familiar type of saving; it's the money that goes into bank accounts or money market funds.

Business saving is less familiar but utterly crucial. Business uses its retained earnings (whatever is not paid out in taxes or dividends) to invest in plant and equipment, or to finance research and development.

As for the government, it hasn't been doing much saving recently—it not only spends whatever it takes in, but then goes out to borrow some more. Nonetheless, the government

SAVINGS

does channel funds into investment goods like highways and bridges.

At the personal level, economists like to distinguish different motives for saving. For instance, people can save for retirement, or against an unexpected occurrence.

As for the exact proportion of income saved, that can be a function of several things. Generally, people with higher incomes save more. (Common sense: Poorer people must spend a higher proportion of their income just to house, clothe, and feed themselves.)

Beyond the level of income, the amount saved can also depend on inflation: The higher the rate, the less incentive to save, since inflation will erode the value of savings. It can also depend on the return to saving; higher interest rates make saving more attractive. Too, tax policy determines incentives to save in the United States; for example, some forms of saving, like Individual Retirement Accounts, are tax-advantaged. And, finally, people who feel that their future prospects are solid will save less than those who are afraid, for example, of losing their jobs.

Say's Law
Named after the French economist, Jean Baptiste Say, who enunciated it in 1803.

Say's Law has been popularly translated into "supply creates its own demand." There are lots of interpretations of Say's law, but the only sensible one is that markets tend toward equilibrium.

Schumpeter, Joseph (1883–1950)
Austrian economist who emigrated to the United States in 1932. Schumpeter wrote about many topics on economics: business cycles, trade cycles, history, and development. He is best known, however, for his work on the importance of entrepreneurs and innovation for economic growth.

Seasonal Adjustment

Fiddling with data that are collected over time.

Consider the statistics on unemployment. Some unemployment is seasonal—there's not much demand for construction workers during the winter, for example. But policymakers, and others, will want to know if a sudden surge in the number of unemployed in December is a natural event or a warning sign that the economy is heading for a recession.

Thus, the figures for unemployment in December will be adjusted—an estimate of how many unemployed are just temporarily jobless construction workers will be subtracted from the total. In other words, eliminating seasonal factors from the data will give a truer picture of the unemployment situation.

SEASONAL ADJUSTMENT

Service Economy

A recently fashionable term used to describe the U.S. economy. Roughly speaking, a service job is anything that isn't a manufacturing job; a secretary working in an auto factory has a service job, while an assembly-line worker in that same factory has a manufacturing job. A service job can be anything from a banker jetting around the world in search of clients to a janitor mopping the basement floor.

The mere existence of the term is a reflection of some broad changes which have been under way during the past three decades. A small but telling example: In 1950, jobs in the service sector accounted for less than half of total employment. Since then, service jobs have grown to somewhere over 75 percent of employment, and the service sector itself accounts for some 70 percent of gross national product.

SERVICE ECONOMY

What happened? Part of the switchover was a natural out-growth of the expansion of business production through the consolidation of firms. And part was an outgrowth of busi-ness expansion from local to national to international reach. Both these developments required an expansion in support services—the communications, transportation, financing, and information processing necessary to keep business flow-ing along at a reasonable clip.

Another contributing factor was the entry of women into the labor force in the sixties. Once these women left home, they needed to purchase the services that they had been pro-viding "for free." Hence the growth in child care and fast-food services, along with fancier stuff like personal shoppers and financial planners. So, too, the explosion in health ser-vices is due to the fact that Americans are living longer and richer, and requiring everything from nursing homes to per-sonal fitness trainers.

Until the 1980–82 recession, an economy that was strong in services but weak in manufacturing was thought to be, somehow, effete. Manufacturing meant industrial might; it was the muscle that built the United States into a world power. But the staying power of jobs and output in the ser-vice industry—power that kept generating jobs and income throughout a recession that saw manufacturing take a hosing —has started to reverse that prejudice.

Indeed, a more reasonable view of the service industry is that it can facilitate manufacturing. What good is a yardful of widgets if the widget maker can't use telephones to sell them, trucks, trains, and airplanes to transport them, and comput-ers to keep track of all those activities.

Short Run
Any period of time too short to allow a firm to change or vary all its factors of production.

Say that Framistan International discovers that its orders outpace its productive capacity. And say that it will take

Framistan International two years to expand its production capacity by building a new factory. Then, everything that Framistan does during those two years—like hiring more workers—consists of things done in the short run.

For some economists, short run is any time period where there is a disturbance in equilibrium. Say that the actual price of framistans is above the market price because demand is very strong. Economists would describe that situation as a disequilibrium in the short run because, by definition, equilibrium is what must happen in the long run. See LONG RUN.

Smith, Adam (1723–1790)

A Scottish philosopher and probably the founder of modern economics. In his classic, *The Wealth of Nations* (1776), Smith proposed a major departure from the practice of letting politicians and kings—or queens—fiddle with the market. Indeed, Smith tut-tutted over a seemingly endless amount of follies and fiascos that had resulted from state interference in the economy.

Smith suggested that the way to promote national wealth was to allow individuals to pursue their own economic interests in a free market. The process by which behavior aimed at selfish gain results in the general benefit is described by Smith's enduring metaphor the "invisible hand." Other terms associated with Smith are "economic laissez-faire" and "free enterprise." No accident, then, that the burgeoning bourgeoisie of the next century—and the established bourgeoisie beyond—cleaved to Smith's celebration of their importance to national health and well-being. See INVISIBLE HAND.

Social Benefit; Social Cost

Something that affects everybody, not just specific individuals or firms. A social benefit like clean air improves the welfare of society, while a social cost like pollution detracts from it. Social costs or benefits are often called externalities.

Socialism

An economic (and political) system where most property is publicly, not privately, owned.

Under socialism there is no market in the sense that prices are set by the government. And there is no profit motive in the sense that profits belong to everyone (income is distributed by the government). Economists call socialism a command, or planned, economy because it is necessary for large government bureaucracies to command, or plan, resource allocation. See PLANNED ECONOMY.

Stagflation

Term popular in the 1970s to describe a situation with both high rates of inflation and low, or negative, rates of economic growth.

Stagflation mystifies some economists because, prior to the 1970s, inflation was thought to be the result of red-hot economic growth. Economists assumed that inflation happened when demand outstripped the economy's capacity to produce —certainly not when the economic activity was slack and a lot of workers were unemployed. During the 1970s, however, inflation zoomed while the economy languished and unemployment stayed stubbornly high.

The usual explanation for stagflation says that after a period of high inflation people learn to expect still more inflation. Thus, price increases are built into contracts—like labor agreements—which, in turn, push prices further up.

The policy dilemma here is that if the government tries to slow inflation by clamping down on spending and money growth, the immediate effect is that the economy contracts, slowing the growth of output, but labor and raw material prices will continue to increase. Simply put, government policy will cause a recession without taming inflation. On the other hand, if the government tries to match the growth in wages and costs with expansionary fiscal and monetary policies, it will create more inflation.

Subsidy

Money that is paid by the government to keep goods or services "affordable."

Usually subsidies take the form of payments to the producers. The money paid to producers raises their income above what it would otherwise have been—they are quite literally paid to produce—with the hope that higher production will lower prices. If all goes according to plan, the subsidy is also a subsidy for the consumers, who, of course, pay less than they otherwise would have. Public transit is a favorite form of government subsidy.

Supply

Shorthand expression for the relationship between the willingness to sell something and its price. It's hard to underestimate the importance of supply for economic analysis. Supply and its twin, demand, are the foundations on which modern microeconomics rests.

Supply has a precise, mathematical meaning for economists. And learning that exact meaning requires digesting several chapters in an introductory textbook, chowing down concepts like iso-quants, production functions, variable and fixed costs, and—for the sophisticated—understanding integral calculus. Scholarly application notwithstanding, supply can also be understood in one, commonsense sentence: Producers will offer more of their product when its price rises, and less of it when its price falls.

The relationship between the quantity supplied and its price is best expressed graphically by something called a supply curve. Consider a supply curve showing how many framistans a framistan producer can supply over a year. Possible prices per framistan can be put on the vertical axis (from $1000 to $4000), the number of framistans on the horizontal axis (from zero to 150).

The supply curve consists of plotting how many framistans the producer will supply to the market at each price; say, for

example, at $1000, the producer will supply 50 framistans, and at $2000, the producer will supply 100. When the points showing how many framistans will be offered at each price are connected by a line, that line will slope upward from left to right. That is, there is a positive relationship between the quantity supplied and the price; as prices rise, quantity increases.

The line's upward direction, what economists call its positive slope, is due to the so-called law of supply. (When properly drawn, the line is actually a curve whose slope depends on the elasticity of supply.)

What's behind the law of supply—other than the notion that people will be willing to offer more of a product the higher its price? For economists, the answer has to do with the cost structure of the firm. Costs come in several guises: There are labor costs, materials costs, equipment costs, capital costs, and so on and so forth. When all these costs are added together, it's possible to figure out the cost of producing one single framistan (or 10,000, for that matter).

Economists focus on something called marginal cost—what it costs to produce an extra unit. When the marginal cost of one more framistan is below its selling price, then it pays to produce that framistan. Why? Because money can be made by selling it. And, of course, if the marginal cost is above the market price, then it doesn't pay to produce more framistans. Simply put, the marginal cost curve—how much extra each additional framistan costs to produce—is the supply curve for the firm. As the prices rise, it pays for a firm to produce more framistans because it can produce farther up on its marginal cost curve. (The more general explanation for the positive slope is the law of diminishing returns—as returns go down, marginal costs go up.)

When all the individual supply curves for framistan producers are added together, the result is a supply curve for the entire framistan industry.

Despite the fact that this explanation seems very precise, almost static, supply curves can be slippery things. Both the

market supply curve, as well as the curve for an individual firm, can shift around. Say, for example, the price of raw stanislaws, the chief ingredient in framistans, falls. That will lower the cost of producing framistans (including the marginal cost) so that the whole supply curve will fall—or, as economists say, shift to the right. Thus, instead of the old relationship where, at the $2000 price, 100 framistans will be supplied, now 150 framistans can be offered. See DEMAND; ELASTICITY; MARGINAL ANALYSIS; SUPPLY AND DEMAND.

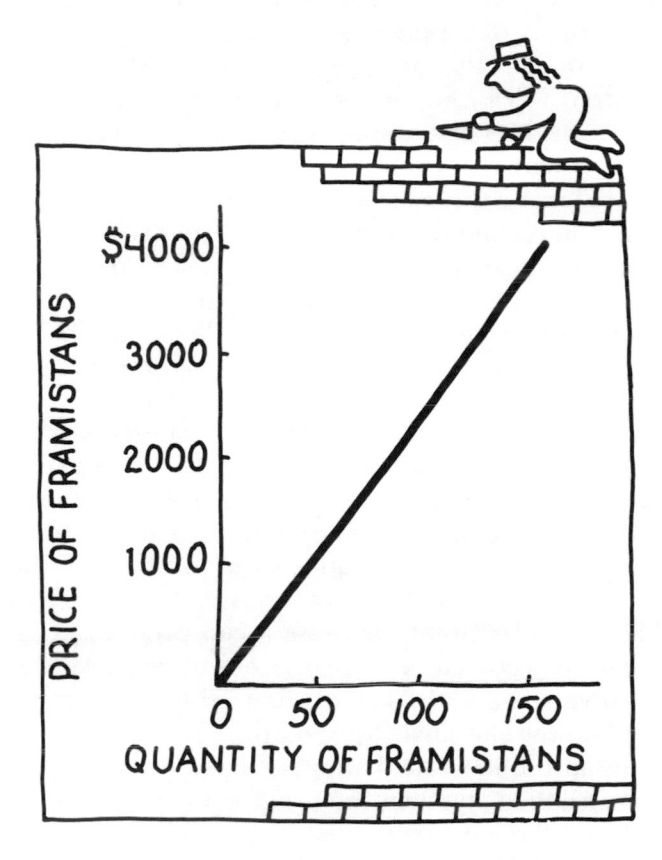

SUPPLY

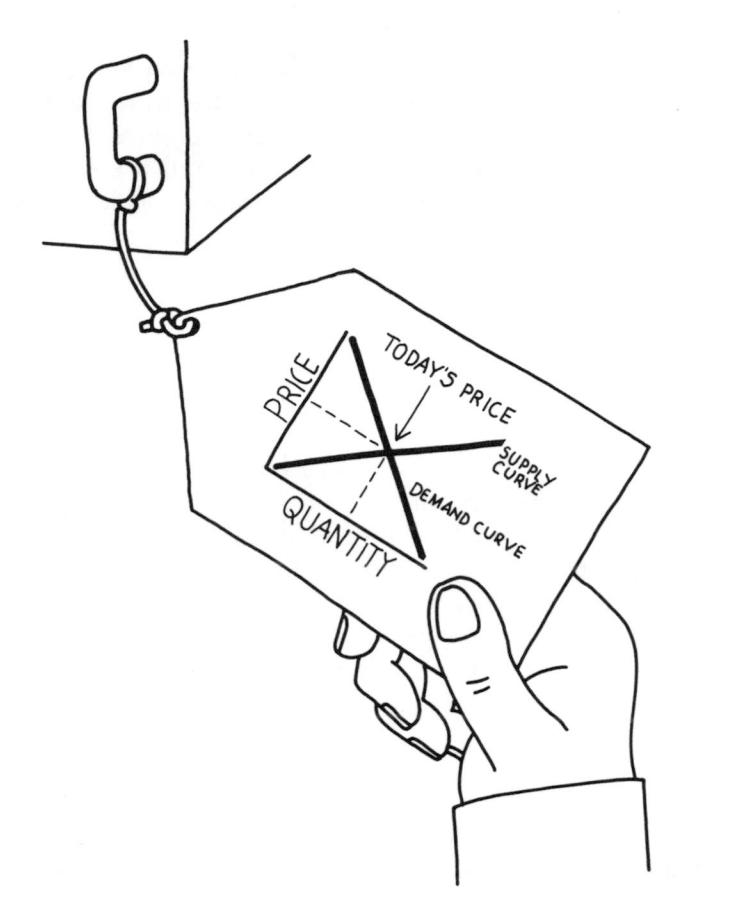

SUPPLY AND DEMAND

Supply and Demand

Term capturing the way prices are set in a free market. This isn't fair, but if you read the Demand and the Supply entries, this one will be a snap to grasp.

Take, on the one hand, the demand curve—that graphical expression of how many framistans will be bought at any particular price. Then take the supply curve—that graphical

expression of how many framistans will be offered for sale at any particular price. When these two curves are drawn on the same graph they will, in all likelihood, intersect since the demand curve slopes down from left to right and the supply curve slopes up from left to right. The spot at which they intersect—when supply and demand meet—determines the price of the framistan. See DEMAND, SUPPLY.

Supply Shock
A sudden disturbance in the supply of some good or service.

The most dramatic kinds of supply shocks happen in the commodity markets. Bad weather in Brazil can deck the coffee crop, for example, and the sudden decrease in the supply of coffee will send prices skyrocketing. Conversely, good weather might produce a giant crop of coffee beans, and the price of coffee will plunge.

The most disruptive supply shock in recent times was the oil price shocks that happened when OPEC organized a cartel that caused the price of oil to zoom ever upward through the 1970s.

Supply-Side Economics
Economics that emphasizes stimulating supply rather than demand.

Demand-side economics and the policies that flowed from it were hot during the sixties and seventies. Government concentrated on pumping up the demand for goods and services. That translated into higher and higher taxes so that government could have larger and larger revenues so it could have more and more to spend. The results of these policies, among other things, were high rates of inflation, stubborn budget deficits, and—eventually—low rates of economic growth.

Supply-side economics, which became the rage in the late seventies, takes an entirely opposite approach. Here, govern-

ment concentrates on increasing supply, or production, through its tax policies. The centerpiece of supply-side economics is the call for cuts in marginal tax rates. It's a simple notion: If you get to keep a larger portion of each extra dollar you earn, you will work more. This incentive effect holds true, too, for saving and investing, since the after-tax return on a dollar saved will go up when the tax rate does down. Thus, supply-side policies are designed to encourage work, saving, and investment.

Further, supply-siders disagree with demand-side orthodoxy over the impact of federal budget deficits. Supply-siders argue that budget deficits, by themselves, are not inflationary; they only become so when the Federal Reserve decides to print money to pay for the borrowing.

While supply-side arguments seem like good common sense, their actual effect is a matter of some dispute. Many economists feel that tax rates in the United States are not high enough to discourage work, saving, and investment; they question whether cutting marginal rates will call forth the productive burst in economic growth claimed by supply-siders. See REAGANOMICS; TAXES.

Tariffs

Taxes imposed on imports.

Tariffs are a favorite tool for protectionists because, unlike quotas and subsidies, they are taxes—they raise revenue for the government. Further, tariffs are perceived as punishment aimed at foreigners; after all, they must pay the tariff.

Nonetheless, tariffs result in the same old protectionist mess. Even though a tariff will cause foreigners to sell less, the tariff becomes embodied in higher prices, regardless of whether the good is foreign-made or domestically produced. Thus, consumers lose (prices go up), producers win (they can raise their prices), the economy suffers (inflationary pressures are created, and diminished competition causes inefficiencies). See PROTECTIONISM; QUOTA; SUBSIDY.

Tax Equity

What designers of tax systems try to achieve.

The notion behind tax equity is that while paying taxes is not pleasant, at least it should be fair. That's easy to say. But there are two aspects to equity—horizontal and vertical. Horizontal means that everybody with the same amount of income should pay the same amount in taxes; vertical means that people with more income should pay more taxes. Unfortunately, under the current income tax system, with its hundreds of loopholes, not only do people with identical incomes pay wildly differing tax bills, but some rich people pay less than some poor people.

Tax Expenditure

Tax laws that lower the tax burden by granting deductions, exemptions, or credits. The reason such legal loopholes are called expenditures is because the government "spends" the revenue by not collecting it in the first place.

Tax Incidence

What befalls the person holding the hot potato when time is called.

Tax burdens can be shifted around. Say you must pay a tax on the bananas you sell. If you simply add it to the price of the bananas, then even though you pay the tax money to the government, the incidence of the tax falls on the buyer.

TAX INCIDENCE

Taxes

No jokes about death and taxes here. Taxes are a serious and complicated matter. Strictly speaking, the purpose of taxing things—earnings or goods or services—is to raise revenue so that the taxing authority can perform certain duties. The taxing authority that looms largest is the federal government. And, presumably, the duties performed by the federal government are commonly agreed-upon functions, from providing for defense and disabled citizens to aiding farmers and small businesses.

Currently, the major tax is one on earnings. Most people pay two types of federal taxes: a straight percent of payroll earnings for the Social Security tax and a progressive percent of earnings for the income tax. "Progressive" means that tax

rates increase with income; it is based on the notion that those who can afford to pay more, should.

So far, so good. But now for the complications. The mere existence of a tax distorts economic behavior. A tax tacked on to the price of a bunch of bananas, for example, means the bananas cost more money. All other things being equal—as economists are fond of saying—people would buy fewer taxed bananas than they would untaxed bananas. Thus, a banana tax can be a powerful way to discourage banana buying.

Ditto for a tax on earnings, particularly given the system of marginal tax rates where each extra dollar of income earned is subject to a higher tax rate. People have less incentive to earn that extra dollar because they will have to pay more of it in taxes or, put another way, they will be allowed to keep less of it. This is known as the incentive effect and—brother—it can be wicked.

So wicked, in fact, that some people, faced with high marginal tax rates, reduce their work effort rather than "working for Uncle Sam." Others are driven into what's politely called tax avoidance: At best, they don't declare some portion of their income or, at worst, they don't file any tax return. The increasing number of Americans who avoid taxes, along with the negative impact the incentive effect has on work (and on savings, too) has prompted some observers to call for a different system of taxation. See FLAT TAX; VALUE-ADDED TAX.

Terms of Trade

One of those international terms to describe who is winning and losing in the game of trade.

The phrase itself characterizes the changing export-import relationship between nations in real terms. That is, a country experiencing favorable terms of trade is one where the value of its exports rises; thus, it can buy more imports for the same amount of exports.

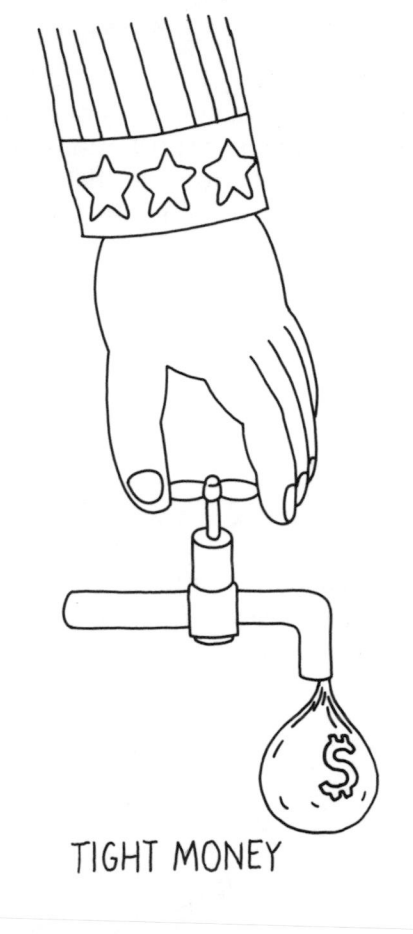

TIGHT MONEY

Tight Money

A completely apt phrase to describe an involved idea. Tight refers to the monetary policy undertaken by the Federal Reserve and money refers to the amount of money in the economy.

When the Fed decides to restrict the availability of credit in the economy, it clamps down on the supply of money. This causes interest rates to rise and borrowing to become unattractive. In other words, tight money means that the Fed has reined in the growth of money.

The opposite of tight money is loose money. See FEDERAL RESERVE SYSTEM; LOOSE MONEY.

Tobin, James (1918–)
Winner of the Nobel prize in Economics in 1981. Tobin has made many outstanding contributions to macroeconomic theory, but he is best known for his defense of Keynesian economic policies. Indeed, Tobin was a member of the Council of Economic Advisers during John F. Kennedy's tenure as president—a veritable festival of Keynesian ideas-put-to-work.

Transfer Payment
A government expenditure for which no service is rendered or product delivered.

Most transfer payments are a form of welfare—Social Security, unemployment benefits, Medicare. The word transfer here really means redistribution. The government collects money from taxpayers and then redistributes it as income.

TRANSFER PAYMENT

Unemployment

The number of workers who aren't working, usually expressed as a percent of the labor force.

Low unemployment—or high employment—is a basic goal of government. Thus, much effort goes into achieving that goal. (The chief platitude is that a fully employed work force makes for healthy economic growth and individual well-being. True. True, too, that a political party that oversees a period of high unemployment usually finds itself unemployed.) Nonetheless, really low unemployment has eluded policymakers in most peacetime years.

There are two kinds of unemployment: frictional and structural. Frictional refers to joblessness that is short term, either because it's seasonal unemployment or because it represents people changing jobs. Thus, stamping out frictional unemployment is impossible and not even desirable.

Structural unemployment refers to joblessness caused by changes in the nature of the economy. The shift out of manufacturing and into services, for example, has caused some structural unemployment. Here, again, stamping out structural unemployment is impossible and not even desirable—an economy that cannot adapt to changes will not be a healthy one.

Beyond these two kinds of unemployment, which set a floor below which unemployment cannot drop, some government policies actually increase joblessness. The minimum-wage law, for example, by setting an above-market wage, means that employers can't afford to hire as many workers as they want. Thus, people who are willing to work for the market wage cannot find jobs. They are involuntarily unemployed. Unemployment compensation, on the other hand, creates voluntary unemployment. Paying people for not working—especially when they are paid more for not working than they would be for working—creates a strong incentive for them to remain jobless.

One final note: For a while policymakers thought they had discovered a trade-off between rates of inflation and the level

of unemployment. This relationship, called the Phillips curve, suggested that a high inflation economy would yield low unemployment and vice versa. Unfortunately, however, the relationship has proved a lot more slippery and less sound than it seemed when A. W. H. Phillips published his findings in 1958.

UNEMPLOYMENT

UNION

Union

Organization of workers that—depending on one's bias—either protects its members against exploitation by owners or management, or exploits nonmembers by raising wages above the market price, which, in turn, restricts employment.

No question about it, unions increase individual economic clout. But they do so by limiting the supply of labor. Some craft unions, for example, hold membership down by requiring large initiation fees, long apprenticeships, or strict licensing procedures. And industrial unions bargain for high wages that can make it too expensive for employers to hire additional workers.

Utility

Workpersonlike word to describe the ineffable—the pleasure or fulfillment that people derive from consuming a good or a service.

Utility is a key notion for microeconomics. Because people receive utility from consuming goods and services, they demand goods and services; and because they demand goods and services, goods and services are produced, and so on and so forth.

For the concept of utility to be, well, useful for economists, however, it can't exist in such a nonquantifiable state. Thus, economists invented the "util"—a unit that measures, loosely, satisfaction. Say, for example, you like potatoes twice as much as broccoli; then you can assign 10 utils to the consumption of one pound of potatoes and 5 to a pound of broccoli. To be really precise, however, you must assign these utils over a period of time, say a week or a month. (After all, it makes a difference in your satisfaction if you consume a pound of potatoes in an hour or over a month.)

Consumption decisions are rarely all-or-nothing propositions, but decisions that are made at the margin. Generally you choose between buying more or less potatoes, not whether you should buy any at all. Accordingly, much of economic analysis deals with marginal, or extra, utility. Take, for example, the marginal utility for the consumption of potatoes consumed over a week. Say you derive 10 utils from one pound, 15 utils from two pounds and 18 utils from three pounds, then your marginal utility is 5 from the second pound and 3 from the third pound.

Note that while your total utility is going up, marginal utility is falling. That is due to what economists call diminishing marginal utility; as you increase your consumption of one good, while holding the consumption of all other goods constant, there comes a point where the pleasure you receive from the last unit will decline. (Indeed, if you consumed ten pounds of potatoes in a week, the last potatoes might be totally pleasureless.)

So far, so good. But so far the explanation is missing a crucial ingredient—price. Even if you get twice as many utils from potatoes as from broccoli, you might still buy a pound of broccoli if potatoes are more than twice as expensive as broccoli. Of course: why spend extra money on something that won't give you the full extra pleasure?

Since most people are constrained by limited budgets, consumption decisions are said to be based on marginal utility per dollar. Economists assume that people maximize their satisfaction and pleasure by trying to consume a mixed basket of things such that the marginal utility per dollar of potatoes is equal to the marginal utility per dollar of broccoli, and so on.

Don't, however, be deceived by how orderly all this sounds. As an idea, utility is very slippery, and as an actuality, the utility derived from goods and services for any one person is always changing. See MARGINAL ANALYSIS.

Vv

Value Added; Value-Added Tax

The difference between the value of a firm's output and the value of the input purchased in order to produce. It represents, quite literally, the value the firm has added.

The relative simplicity of the arithmetic that goes into figuring value added makes it a juicy target for taxation. Hence, the popularity of a value-added tax, especially in Europe.

A value-added tax is levied at each stage of production. Thus, the farmer who buys seeds and fertilizer pays a tax on the difference between the cost of those raw materials and the value of the wheat she sells to the miller. She will, of course, pass the tax on to the miller by including the tax amount in the price of the wheat. The miller who turns the wheat into flour must pay a tax on the difference between the value of the wheat and the flour, and he, too, will pass the tax on to the baker in his flour prices. The baker, of course, will pass his tax on to his buyer, the supermarket, by including the tax amount in the price of his bread. In the end, it's the consumer who is thought to pay the whole burden of the tax, which will be embodied in the price of the bread at the supermarket's checkout counter.

VALUE - ADDED TAX

Variable Cost

Known informally as operating cost. Any cost that varies with a firm's production. (When a firm isn't producing anything, it's variable cost is zero.) Examples are labor costs, raw materials costs. See FIXED COST.

Ww

Welfare State

Exaggerated way of indicating the mechanism by which a mostly market economy fulfills its responsibility to its less fortunate citizens.

In a welfare state, the government assumes the major role in redistributing income from richer to poorer. There are two main mechanisms of redistribution. The state can use its tax revenues to pay for services or goods for the poor, and/or the state can base its system of taxation on the principle that richer citizens should pay more than poorer ones.

A welfare state is not the same as socialism. In a welfare state, the means of production are still, by and large, privately owned.

Windfall Profits

Rather imprecise term to describe any unexpected increase in income received by individuals or firms.

Usually windfall profits accrue to owners of natural resources that are very sensitive to swings in supply-and-demand conditions. A frost on the coffee crop, for example, will produce windfall gains to any plantation owner lucky enough to have escaped the bad weather.

Despite the fact that windfall profits are fortuitous, they seem to create disapproval and, thus, the impulse to tax. Back in the 1970s, for example, the stunning run-up in oil prices led the Carter administration to impose a windfall-profits tax on oil companies. See PROFIT.

World Bank

More formally known as the International Bank for Reconstruction and Development, the World Bank was created in 1944 as part of the forty-five-nation accord reached at Bretton Woods, New Hampshire. The bank's funds come from its member countries (with richer countries making the largest

WINDFALL PROFITS

contributions), from sales of its own bonds, and from earnings on its loan portfolio.

Like any other bank, the World Bank makes loans. Unlike other banks, however, the World Bank is specially charged with making loans to countries that can't get financing from private sources. In practice, that means the bank lends money to a lot of less developed countries for a lot of risky investment projects.

Supporters of the bank argue that poor countries need capital to develop; thus, the bank performs a necessary and useful function. Critics reply that the bank underwrites dubious undertakings that, once created, often need subsidies to operate; thus, the bank leaves debtor countries less able than ever to operate in the world economy.

ZERO-SUM GAME

Zero-Sum Game

An element of game theory where one person's gain is another person's loss. Thus, the winner's gains equal the loser's losses so that the total sum of gains and losses is zero. Less clumsily said: If you get $10 richer, then I must get $10 poorer. The idea has become popular with economists because, it seems, most problems in public policy—like taxation—involve zero-sum solutions. See GAME THEORY.

ACKNOWLEDGMENTS

Thank you to those who read parts of the manuscript: Tim Ferguson, Robert Barro, Chris Byron, Jim Flanigan, and Shelly Zalaznick.

And gratitude to my editors at *Vogue* magazine, who were willing to take a chance on a column written by an economist; to my friends who exhibited deep and abiding faith: Susan Chace, Deirdre Rosenberg, Peter Martin, Christopher Norwood, Bob Wilson, Ina Finkel, John Nanney, Tatiana Pouschine, Del and Adelaide Hubbell; and to my editor, Elaine Pfefferblit, and my agent, Ginger Barber, who were nothing if not cheerful throughout.

But most of all, my thanks go to the two editors who taught me that thinking like an economist doesn't have to mean writing like one, Bob Bartley and Jim Michaels.

ABOUT THE AUTHOR

Susan Lee is a senior editor of *Forbes* magazine, in which her column appears regularly. She has been a member of the editorial board of *The Wall Street Journal* and a columnist on public policy issues for *Vogue*. From 1976 to 1980, she taught economics at Columbia University, from which she holds a Ph.D. in economic history. Her articles and reviews have appeared in *The New York Times*, *Barron's*, *The Washington Post*, and *The New Republic*. She lives in New York City.

PublicAffairs is a publishing house founded in 1997. It is a tribute to the standards, values, and flair of three persons who have served as mentors to countless reporters, writers, editors, and book people of all kinds, including me.

I. F. STONE, proprietor of *I. F. Stone's Weekly*, combined a commitment to the First Amendment with entrepreneurial zeal and reporting skill and became one of the great independent journalists in American history. At the age of eighty, Izzy published *The Trial of Socrates*, which was a national bestseller. He wrote the book after he taught himself ancient Greek.

BENJAMIN C. BRADLEE was for nearly thirty years the charismatic editorial leader of *The Washington Post*. It was Ben who gave the *Post* the range and courage to pursue such historic issues as Watergate. He supported his reporters with a tenacity that made them fearless and it is no accident that so many became authors of influential, best-selling books.

ROBERT L. BERNSTEIN, the chief executive of Random House for more than a quarter century, guided one of the nation's premier publishing houses. Bob was personally responsible for many books of political dissent and argument that challenged tyranny around the globe. He is also the founder and longtime chair of Human Rights Watch, one of the most respected human rights organizations in the world.

·　　·　　·

For fifty years, the banner of Public Affairs Press was carried by its owner Morris B. Schnapper, who published Gandhi, Nasser, Toynbee, Truman, and about 1,500 other authors. In 1983, Schnapper was described by *The Washington Post* as "a redoubtable gadfly." His legacy will endure in the books to come.

Peter Osnos, *Founder*

HEMANT TANEJA is a managing director at General Catalyst, a prominent venture capital firm. Unscaling is the central investment philosophy driving his work with groundbreaking companies such as Stripe, Snap, Airbnb, and Warby Parker. Hemant is the cofounder of Advanced Energy Economy, an organization focused on transforming energy policy in America, and is a board member of Khan Academy, a nonprofit educational organization. Additionally, he serves on the Stanford School of Medicine Board of Fellows. Hemant teaches a course on AI, entrepreneurship, and society at Stanford University and first published on the unscaling phenomenon in the *Harvard Business Review*. He holds five degrees from MIT.